Technology in World Civilization

In memory
of my father
Leslie Pacey
and of
his vision:

天 下 一 家

'(all the world) under heaven one household'

Technology in World Civilization

A Thousand-Year History

Arnold Pacey

The MIT Press
Cambridge, Massachusetts

First MIT Press edition, 1990
© 1990 Arnold Pacey 2 00 /

Printed and bound in Great Britain

Library of Congress Cataloging-in-Publication Data
Pacey, Arnold.
    Technology in world civilization : a thousand-year history /
Arnold Pacey.
        p.   cm.
    Includes bibliographical references.
    ISBN 0–262–16117–6
    1. Technology—History.   2. Technology transfer—History.
I. Title.
T15.P353   1990
609—dc20   89–12801
                                                        CIP

Contents

Preface

Historical themes in technology

One puzzle which challenges anyone who attempts to study the development of technology over the whole world is that every now and then, similar inventions appear in different parts of the world almost simultaneously (for example, the spinning wheel), or at different dates but with no obvious connection (printing, first in China, and six centuries later in Europe). Sometimes these are genuinely independent inventions. Sometimes there is clear evidence of information spreading from one place to another (as with paper-making). But quite often, the most important factor was that the achievements of one society stimulated people elsewhere to make different but related inventions.

Stimulus effects have been important in other ways as well. Gunpowder and some basic guns were invented in China, but when information about them reached Europe, it provoked a new and more formidable invention, the cannon. Transistors (and many other electronic devices) were invented in the United States, but led to the development of new kinds of consumer product in Japan.

Discussion of similar developments as they occur today has tended to concentrate on the situation where information and often equipment is taken directly from one society to another. In the current jargon, this is a 'transfer of technology'. In most such discussion, however, no allowance is made for the stimulus which can arise from the spread of knowledge, or even from the mere rumour of an unfamiliar technique. An example is offered by Indian textile technology, which had a profound influence in Britain during the industrial revolution even though there were few 'transfers' of technology. Just the knowledge that Indians could spin fine cotton yarns, weave delicate fabrics, and dye them with bright and fast colours stimulated British inventors to devise new ways of achieving the same results.

Typically, the interactions which lead to such results are like a conversation in which incomplete information sparks new ideas and what we can call 'responsive inventions'. Sometimes it is like a dialogue or dialectic in which recipients of a new body of knowledge and technique 'interrogate' it on the basis of their own experience and knowledge of local conditions. In these instances, the initial 'transfer of technology' itself is only the first stage in a larger process.

Failure to appreciate this dialectic in technology can lead to a rather restricted view of the history of invention, but more seriously, it has led to policy failures in the modern world. Programmes designed to encourage transfer of technology from industrial nations to 'less developed' countries have often been frustrated because they have not allowed for responsive invention in the countries concerned. The transferred technology has effectively been *imposed*. It may function for a while, especially if engineers and managers from the country of origin remain in charge. Or local people may be forced to reconstruct their lives around it. But the experience is often that the technology does not function well and is ultimately abandoned. Efforts to introduce tractors, water-pumps, sewage works and factories have gone through this cycle in many African countries, and sometimes in Asia and South America also. The way to avoid such negative results is to introduce the new technology in a more flexible form to allow for a dialogue which may lead to modifications, possibly in equipment, but more especially in social arrangements affecting its use.

However, while dialogue or dialectic in technology is the main theme in this book, there are many other themes which a history of technological change ought to address. For example, in accounting for the way invention has sometimes been stimulated, we must recognize that machines and other artefacts (notably guns) may carry with them a provocative symbolism. By contrast, in understanding the caution with which other inventions are approached, we need to appreciate that when the basic survival needs of a people are at stake, especially their food supplies, they tend to take a low-risk approach to suggested innovations. Some of the questions which arise about symbolism and 'survival technology' are mentioned in later pages, but without a full discussion.

Other themes dealt with only briefly because of lack of space include the *institutions* within which technology develops, and the contrasting influences on innovation of governmental, military, commercial and other kinds of organization. Almost nothing is said, either, of the *beliefs* which different societies bring to their thinking about technology, such as the beliefs, once widely held, that some techniques or machines were capable of tapping unnatural or 'magical' forces. Nor is there any discussion of the empirical knowledge used by craftworkers and its possible status as 'proto-science'. Some of these topics are dealt with more fully in another of my books, *The Maze of Ingenuity*, which covers the same historical period as this volume, but in relation to developments wholly within

western cultures. Stimulus from non-western cultures was always important in Europe, but direct transfers of technology were fewer than often
claimed. European technologists had a notably imaginative approach for
which the symbolism of machines and structures and ideas about magic
were sometimes important.

In discussing more practical aspects of these themes, we need to beware
of the misunderstandings which can arise from the widespread habit of
equating the word 'technology' with 'engineering'. For example, we can
too easily underestimate a society in a dry tropical country which uses a
subtle understanding of ecological relationships to grow crops without
heavy irrigation works. Lack of irrigation engineering in such a country
does not imply lack of technology, but sometimes quite the reverse. It is
worth adding that when 'engineering' is mentioned here, it refers to
techniques for constructing earthworks, machines, canals, bridges and
building structures. It does not refer to either 'civil' or 'mechanical'
engineering in the modern sense but to whatever combination of these
skills existed in the society under discussion (which often included
architecture also).

One other definition which may be necessary is that the word 'industry'
refers here to any organized concentration of manufacturing or metal
smelting, even if the context is what might sometimes be described as a
'pre-industrial' society. Often the term should be understood to mean a
'handicraft industry' with little mechanization, but some large-scale, full-
time industry quite separate from agriculture has existed in places for a
surprisingly long time.

Another convention used in the book is that all measurements have
been converted to metric units except that weights given in tons in source
material have been quoted unaltered. The differences between metric
tons, long (imperial) tons, and short tons are not great (1 metric ton or
tonne = 0.984 long tons = 1.102 short tons). There is uncertainty about
which unit some authors are using, and since many figures are already
rounded to the nearest thousand tons, conversion would in any case
make little difference.

Pinyin spelling has been adopted for Chinese names, and comparable
modern spellings in their simplest form for other non-European words.
Users of Joseph Needham's monumental work on Chinese science and
technology, which is inevitably much quoted, can convert from pinyin to the
spelling convention he favours via tables at the end of his more recent
volumes. With regard to names of countries, 'Iran' and 'Iraq' are used here
to denote geographical areas, not political entities, which is consistent with
the way these words were used during most of the period discussed.

Given the immensity of the subjects discussed here and a lack of
research resources, all that has been attempted is a broad perspective

based on reference material in English. But broad perspectives are often helpful for the 'general reader', and are becoming increasingly important in this world of specialists for those undertaking more detailed studies – in this case, in the history of technology, science policy, or development studies. The limitations of the research are reflected by the notes at the end of the book, which have been kept relatively simple. Apart from the works referred to there, however, a good deal has been learned from a variety of museums and from the comments and practical help of a number of people. Individuals to whom I am conscious of specific debts include: Lisa Barker, Tony Barker, Dinie Blake, Robert Chambers, Randolf Cooper, Peter Ewell, David Farrar, John Fassnidge, Anil Gupta, Donald Hill, Penni Jones, Philip Pacey, Megan Parry, Jerry Ravetz, Robert Rhoades, Stuart Smith, Barrie Trinder, Charles Webster and Zheng Yongning.

Among libraries which have taken trouble to obtain books for me and organizations which have provided information – sometimes small but vital details – I am conscious of specific debts to: Bankfield Museum (Halifax), Beijing Ancient Observatory (China), Hartlepool Ship Restoration Company (and especially Walter Brownlee), Ilkley Public Library, Ironbridge Gorge Museum (and John Powell), Lancashire Polytechnic Library (Preston), Leeds University Library, National Army Museum (London), National Maritime Museum (Greenwich), Sichuan Brocade Factory (Chengdu, China) and the Victoria and Albert Museum (especially D. A. Swallow and Betty Tyers).

Apart from the specific and practical help of these people and institutions, I have to acknowledge two debts of a more general kind. One is to the Open University, on whose history of technology course (code A281) I was working as a tutor while this book was being written. This has inevitably influenced my approach to nineteenth-century technology. The other debt is to my father, Leslie Pacey, who has been very present to me during the writing of the book, not least because circumstances associated with his last illness meant that the early chapters were drafted in his sickroom. This experience reinforced my sense of sharing his view of the world, which he expressed in work that once took him to China, and which he found encapsulated in the old saying quoted on a previous page. The first half of this, 'all under heaven', is an archaic phrase meaning 'the civilized world', usually identified with China as distinct from the bleak lands to the north or countries overseas. However, the same phrase can today be taken to mean 'all the world', and the saying as my father understood it refers to the whole world as one household, and humankind as one family. That sets the context for the dialogues described here, and for the environmental concerns of the present generation.

Finally, I am much indebted to four artists who have contributed illustrations, namely Hazel Cotterell, Clare Hemstock, John Nellist, and my mother Mildred Pacey. Ironbridge Gorge Museum kindly allowed me to use illustrative material from their collections. The following publishers have also given permission for illustrations which originally appeared elsewhere to be reproduced here, or to be used in substantially modified form:

Calderdale Museums, Halifax (Figure 13)
Cambridge University Press (Figures 6, 8, 9, 14, 15)
Croom Helm, Beckenham, Kent (Figure 12)
Intermediate Technology Publications, London (Figure 44)
International Development Research Centre, Canada (Figure 41)
Phillimore & Co., Ltd., Chichester (Figure 37)

Arnold Pacey
Addingham
January 1989

1 An age of Asian technology, AD 700–1100

A balance in world history

Around a thousand years ago, in several regions of the world, there was a phase of striking invention and technical improvement in agriculture, metal smelting, and some aspects of engineering. These developments were most pronounced in China and in the West Asian countries which had recently come under Islamic rule (including Iran, Iraq and Syria). However, other areas were involved as well, extending even to Europe.

It can be cogently argued that increases in population have often been a spur to technological innovation, especially when more food and other necessities have to be produced from a fixed area of land. This may have been a factor during the period being discussed, for population growth had accelerated in many areas. From AD 700, if not earlier, there is evidence of new cropping patterns in China, West Asia and Europe, as well as developments affecting farm implements, and (except in Europe) improvements in irrigation methods. Such changes were a matter of 'survival technology' related to the basic support of increasing numbers of people.

Quite a different factor stimulating technological change, however, was the trade in luxury goods between China and West Asia. It is clear that the export of Chinese silks was a stimulus to textile industries in the West, and that Chinese paper-making techniques spread as a result of contacts along the trade routes. In the other direction, Iranian windmills became known in China, and eventually led to invention of a different but related type of windmill there. It was as if a dialogue had developed, with stimulus to invent and improve techniques stemming from exchange of artefacts and ideas.

Trade links also meant that commercial prosperity in one region could influence economic conditions in very distant areas, and one historian, W. H. McNeill, argues that the growth of commerce in China during the centuries 'on either side of the year 1000' was such that it 'tipped a critical balance in world history'.[1] McNeil's main case-study of commercial expansion concerns the Chinese iron industry, which manufactured weapons and farm tools, and developed specialized qualities of iron for

1

particular purposes. For example, cast iron is normally regarded as an unpromising material for making either bells or plough-shares, but was successfully used in both applications. Especially good quality control was necessary when large temple bells were made from 'white' cast iron. The earliest of these bells dates from 1079, and many others were made, chiefly at foundries in Shanxi province.

The most rapid expansion in the Chinese iron industry was occurring further to the east, however, in an area straddling the modern boundary between the provinces of Hebei and Henan. Not only was there iron ore here, but there was also coal. Wood or charcoal fuel for furnaces was becoming scarce because of deforestation, and the expansion of the iron industry depended on an increased use of coal, coke and, in Shanxi, anthracite, depending on the type of furnace used.

Figures for iron production in Hebei and adjacent areas[2] show that output increased from a very low level in AD 998 to more than 35,000 tons per year in 1078. The biggest furnaces then in use produced about 2 tons of iron per day, but most were considerably smaller, so this output indicates that very many furnaces were in operation. Large enterprises run by independent entrepreneurs developed, although elsewhere in China iron smelting was usually a part-time occupation taken up by peasants in seasons when agricultural work was slack. Even so, iron production was increasing in other regions as well as in Hebei, and the total output recorded by tax officials rose from 32,500 tons per year in AD 998 to 90,400 tons in 1064, and to 125,000 tons in 1078.

Much of the iron produced in Shanxi province was smelted in crucibles heated in coal-fired furnaces, but it is probable that the rapid expansion in cast iron production between 998 and 1078 depended more on the use of blast furnaces. Judging by somewhat later illustrations, a typical blast furnace was 3 to 4 metres in height, and either barrel-shaped or narrowing markedly at the top, from where it was fed with coke or charcoal, iron ore, and probably some limestone (figure 1).

When a furnace was tapped, white-hot molten metal ran out from the opening near its base. This might be directed straight into moulds to solidify as pig iron, or it could be run into a large square trough where the first steps in converting pig iron into the more malleable wrought iron were carried out.

The air blast required to maintain the chemical reaction in the furnace was typically provided by two enormous hinged bellows, blowing alternately. These were sometimes driven by water-wheels, but were often manually powered, with teams of between four and six men working in shifts. Production figures indicate a large market for iron, and indeed, a big area could be served because the iron-working district was connected by canal to populous regions further south, including Kaifeng, then the capital city (figure 2).

Figure 1 Impression of a blast furnace in the Hebei iron-working region, with four men working the bellows to supply the blast.

The drawing is based on the earliest known picture of a blast furnace in China, dating from 1334, but with shapes and proportions based on other illustrations and discussion of furnace types by Hartwell, Needham, and Wagner.

(Illustration by Clare Hemstock)

Not far to the north of this iron-working region was the so-called Liao Empire. Among its population were many nomadic groups who were formidable warriors and horsemen, and there was frequent raiding across the border. In other respects, the culture and technology of this empire was comparable to that of the heartland of China, ruled by the Song dynasty.

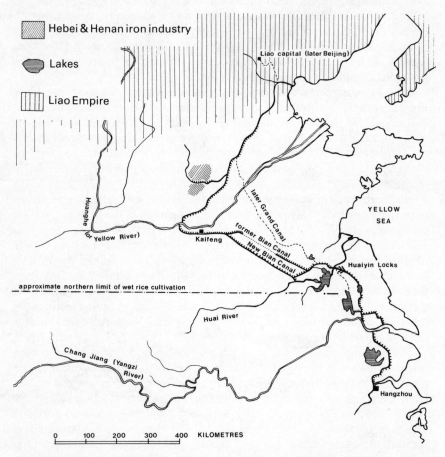

Figure 2   Map of North China showing the relationship of the Hebei and Henan iron industry to the canal system and the Song dynasty capital at Kaifeng.

A pioneer set of five pound locks was built at Huaiyin in AD 983. Elsewhere, changes of level were managed by double slipways or flash locks. The later part of the Grand Canal was constructed at the end of the thirteenth century after Beijing had become the capital of all China.

In response to the Liao threat, Song China built up an enormous army, exceeding one million men by the 1040s. The most important weapon used by the soldiers was the crossbow, but the army also had some incendiary weapons made with gunpowder. Much of the iron produced in the Hebei region went to manufacture military equipment. Government factories were making many thousands of suits of armour each year, and could turn out 16 million iron arrow-heads annually for the crossbow men. However, these military preparations were to some extent in vain. In 1125, another group of nomadic warriors took over Liao territory in the north. These were the Jin (or Jurchen) tartars, and they soon began to invade Song territory. The canals were cut, ironworks were disrupted, and in 1126 Kaifeng, the capital, was evacuated under duress, the government moving south to Hangzhou.

Although the arms industry was the biggest user of iron, the blast furnaces of Hebei supplied many other consumers, some of them Buddhist temples. Here there was enthusiasm for using cast iron in building as well as for bells. Temple roof tiles or even small pagodas were occasionally made of iron. Far more important, however, was the use of iron in ploughs and farm tools. Early Chinese civilization had developed chiefly in the north of the country, in the Yellow River basin. This area has a cool climate and the economy depended on wheat and millet crops rather than rice (see figure 2). But a movement of farmers to the warmer, wetter southern regions is perceptible from AD 700. After 907, this trend was accelerated by raiding in the north. By 1080, instead of the largest part of the Chinese population living in the Yellow River basin, there were more than twice as many people in the south. A smaller area of land there had to support the growing population and at the same time feed the army. The result was an impressive series of innovations in the rice-growing areas, which culminated in very high levels of farm output relative to the land resource available.

The government was active in this and attempted to increase the area under cultivation by starting land reclamation schemes. However, the most vital official project was the introduction of a new variety of rice from Champa, in what is now Vietnam. This was a quick-growing rice which could also be planted early, leaving time after it was harvested for a second crop to be grown. Samples of seed were widely distributed to farmers in South China in 1012, and the crop quickly caught on. One extra advantage was that it could be grown on land where there was an insufficient water supply for ordinary rice.

The introduction of Champa rice came at the culmination of almost two centuries of agricultural development, another aspect of which was innovation in farm implements and their adaptation to rice culture. Better ploughs with iron shares and mould-boards had been introduced before

AD 880. There were also animal-drawn harrows with iron components, new hand-tools for weeding, better sickles, a winnowing machine for use after threshing, and mills for husking and polishing the grain.

Hydraulic engineering

Other developments under way at this time were improvements in the canal system. Boats on the canals were used for transporting grain to supply the capital city and the army, and to move grain levied from farmers and landowners by way of taxation. But with expanding agricultural output, and the need to carry iron from the developing furnaces, canal traffic was inevitably increasing. Grain carryings reached about 400,000 tons per year during the eleventh century.

The Song dynasty capital at Kaifeng was near where the main canals joined the Yellow River. However, after the Mongol conquest of China in the 1260s, the new capital was established at Beijing (Peking), and this made it necessary to build a new waterway – the Grand Canal – to link Beijing with the south.

Many canals were little more than improved river channels or artificial links between rivers. Until the Grand Canal was built there were no canals with a 'summit level' requiring an independent water supply. Most canals were built over fairly flat terrain and changes of level were few; where they did occur, boats were moved from one level to another by hauling them over double slipways. That is, a length of canal would end with a ramp where boats would be hauled out of the water with ropes and windlasses. The top of the ramp led on to a slipway down which the boats could be launched on to another stretch of canal at a different level.

Boats were sometimes damaged on double slipways and their cargoes were vulnerable to theft. Disturbed by this, a government transport commissioner named Jiao Wei-ye in AD 983 instructed that five double slipways on a canal opening onto the Huai River (see figure 2) should be replaced by double sets of lock gates. There was a short length of canal between each pair of gates whose water level could be varied to match either the 'upstream' or the 'downstream' level. Thus the boats could be floated all the way down the five steps from the canal to the river.

These are the earliest *pound locks*, and are notable in being assigned to a named inventor.[3] However, the pound lock was less efficient than the modern chamber lock, in that the short length of canal between the two lock gates was slow to fill, and took a lot of water. Thus double slipways continued to be used in many places, though they were replaced on the New Bian Canal between 1023 and 1031.

Another point to notice about the invention of the pound lock is that, like the introduction of Champa rice, it shows how officials in central or local government could be active innovators. There are several other examples of technological innovation originating within the bureaucracy during the years when the Song dynasty ruled at Kaifeng (AD 960–1126), some related to clocks and some to military equipment. There was also an active commercial sector in the iron-working areas and technological innovations sometimes emerged from Buddhist monasteries, so all things considered, this was an especially creative period for Chinese technology. In 1100, China was undoubtedly the most technically 'advanced' region in the world, particularly with regard to the use of coke in iron smelting, canal transport and farm implements. Bridge design and textile machinery had also been developing rapidly. In all these fields, there were techniques in use in eleventh-century China which had no parallel in Europe until around 1700.

Although there was continuing technological development in China under subsequent dynasties, the balance of social and economic forces which had fostered innovation during the early Song period was disrupted by the invasion of the Jin tartars in 1126, and more seriously by the devastating Mongol conquest just over a century later. Moreover, whilst officials in government services were responsible for several innovations under the Song dynasty, the bureaucracy is more often seen as frustrating technological development in later centuries.

Despite all this, it would be a mistake to view Song China as an island of brilliant innovation in a technologically primitive world. China had active trade links with Iran and the Persian Gulf region, both by the overland 'silk route' and by sea. There were also contacts with Southeast Asia and India. Not only did Chinese exports stimulate local developments in these places, but China was also indebted to them for improvements in agriculture and other technologies. Champa rice and new types of vegetable were introduced into China from the south, and cultivation and processing of cotton had come from India somewhat earlier (around AD 600).

An instructive illustration of how technical ideas were spread is provided by methods for distilling crude petroleum in places where it seeped from the ground, and its use to make incendiary weapons. The military use of this substance to produce 'Greek fire' was initially the work of Byzantine Greeks, perhaps in the seventh century. From them it passed to the Islamic civilization. Then Arab seamen trading from the Persian Gulf to the Malay peninsula and Indonesia introduced the technique in the latter area, where it was quickly exploited because of easy availability of petroleum from surface seepages in Sumatra. The Chinese learned of Greek fire in AD 917 or just before through contacts with Southeast Asia, and were soon using 'fire oil' in new weapons of their own invention.

With these examples in mind, we can perhaps think in terms of a 'technological dialogue' or an 'inventive exchange' between different regions in Asia, through which not only were techniques 'transferred' from one place to another, but also invention was stimulated in response to the transferred technique. With regard to textiles, crop plants and minerals, India's role in the dialogue was especially important. With respect to hydraulic engineering and machines, however, the most significant exchanges were between China and Iran (or the Islamic world in general).

In Iran and Iraq, hydraulic engineering stands out as a branch of technology which was well advanced long before the eleventh century (as was the case in China also). Dams, canals and irrigation works had developed strongly in this region when it was part of the old Persian Empire. There had also been some stimulus from Roman engineering. On one famous occasion, a whole Roman army was captured in battle and was put to work building a dam. This was a way of exploiting the engineering knowledge of the soldiers as well as their labour.

Since that time, however, government throughout the area had been transformed by the advent of Islam. Following the death of Mohammed in AD 632, Islamic armies moved north and east from Arabia, to conquer Syria by 636 and the Persian Empire by 649. Further conquests took in the whole of North Africa, much of Spain, and also Sicily, until Islamic rule extended from the Atlantic to the northwest margins of India.

Since all these lands had relatively dry climates, irrigated agriculture was important wherever there were large concentrations of people to support. However, it is often said that hydraulic engineering made little progress under Islamic rule. The suggestion is that Persian dams and irrigation works were simply repaired, or their designs were adapted to new sites with no advance in technique. Islamic engineers knew about the mechanical devices described by the Roman author Vitruvius and by Greek authors from Alexandria, and it has been argued that the machines they built did not develop beyond this. A more positive interpretation would be that the Islamic civilization adapted the ancient techniques they inherited to serve the needs of a new age, and in particular, that they extended the application of mechanical and hydraulic technology enormously.[4]

One example is provided by an old Persian technique for obtaining a water supply by tunnelling into a hillside, using a series of well shafts for ventilation and access (as illustrated in chapter 5). Such tunnels or *qanats* had previously been used only in Iran and areas close by, but under Islamic rule the technique was adopted in many parts of North

Africa, wherever the terrain was suitable, as far west as Morocco.

In Iraq, there was an urgent need to repair and extend existing irrigation works, particularly after the leading Islamic ruler, the caliph, established his capital at Baghdad in AD 763. Located on the Tigris River, the city was not far downstream from where two major dams diverted water from the river into the Nahwran Canal, which supplied water to irrigate land over an expanse stretching 200 kilometres downstream. This canal was one of the biggest engineering achievements of the Persian Empire and was constructed between AD 530 and 580. Islamic engineers not only repaired the dams and cleared the canal of silt, but built a new dam on a tributary, the Uzaym River, to provide more water (figure 3). This was a masonry dam just over 170 metres long, with a crest 15 metres above the river bed. As in some other dams in the area, the carefully cut stone blocks from which it was made were

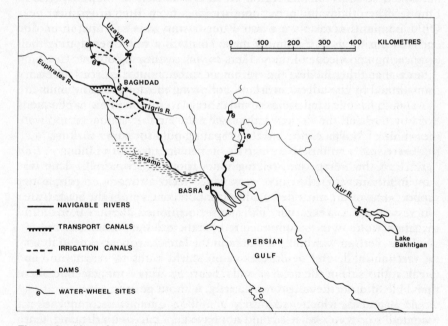

Figure 3   Map of southern Iraq and western Iran showing the location of major works of hydraulic engineering in use during the tenth century. Some were constructed much earlier, under the Persian Empire.

At the water-wheel sites marked, there were often wheels driving mills of different types, and there might have been water-raising wheels also. Many small water power sites are omitted, and the map is generally much simplified.

(Based on data and maps given by Norman Smith, Donald Hill, Hans Wulff, also al-Hassan and Hill.)

joined together by iron dowels. The holes in which the dowels fitted were filled by pouring in molten lead.

Another famous dam, on the Kur River in Iran, was rebuilt about AD 960. Like the other masonry dam mentioned, it was constructed by clamping stone blocks together with iron bars set in lead. Indeed, there may have been an older structure on the site built in the same way. Water released from this dam could irrigate a large area downstream where sugar, rice and cotton were grown.

Irrigation water for such crops was sometimes also provided by using water-raising wheels, some powered by oxen or donkeys, and some driven by the force of the current in the river from which water was being raised. The latter type is illustrated in figure 4. This kind of water-raising device was used in India from a very early date, and had a long history in the later Roman Empire and in China also.

One description of the Kur River dam mentions ten water-wheels in the vicinity driving mills, but another account, written around the year 990, implies that there were also water-raising wheels in operation. The reference to mills could simply mean corn mills, but a new type of mill had been introduced in Iran and Iraq for processing sugar cane. Crushing the cane and then boiling the extract were the main operations necessary for obtaining crystalline sugar. Water-powered cane-crushing mills are known to have existed at Basra and other places. Other new applications of water-wheels developed in the region during the Islamic period were for fulling woollen cloth, and for preparing pulp for paper-making.

There was a remarkable diversity of machines for corn milling in Iran and Iraq, including one striking innovation, the windmill. This was apparently developed in dry areas which lacked streams to turn water mills. Some of the most definite information comes from the Seistan area in eastern Iran, where windmills were mentioned about AD 950, and where a group of mills continued to operate into the present century.

These Persian windmills differ from the later European type in having a vertical shaft, and in depending on shield walls to direct the wind against one side of the array of sails (figure 5). This arrangement made it possible to drive the millstones directly, without gearing.

At Baghdad, which had nearly a million inhabitants, conventional windmills or water-wheels could not have kept pace with demand. Corn milling was carried out by a series of floating water mills on the Tigris River which operated continuously, night and day. At Basra, there were mills driven by tidal flow. In both places, the water-wheels were of the undershot type, driving the millstones through wooden gears.

By contrast, in rural areas, and especially in mountainous regions, there was a smaller type of water mill which had a horizontal, turbine-like wheel, with the millstone mounted on the same shaft, directly above

Figure 4   Water-raising wheel of the type known as a *noria*.
   This is an undershot water-wheel, driven by the flow of water in the stream running underneath it. The jars tied to the rim fill from the stream, and then empty into the trough at the high level.
   (Illustration by Hazel Cotterell)

(see figure 11). By the year 1000, horizontal wheels of this type were used very widely throughout the Eurasian land-mass, from western Europe through Iran to China, and probably also in mountainous regions of North India and Nepal.

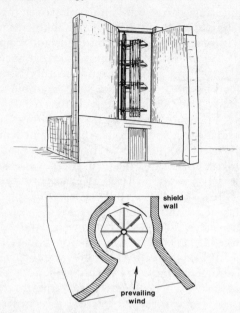

Figure 5   Persian-type windmill showing how shield walls were designed to funnel wind onto the sails, which rotated on a vertical shaft.

This is most clear from the plan (lower illustration). The perspective view shows how the shield walls of later mills formed a towering open structure built of sun-dried bricks with a small room containing the millstones underneath. The earliest mills seem to have been of different design with the millstones supported by the shield walls at the top of the structure.

(Based on photographs of a windmill in Afghanistan by Dick Day.)

Indian Ocean trade[5]

As early as AD 750, Arab shipping was prominent throughout the Indian Ocean, and merchants from the Persian Gulf even made the long voyage to China. As time went on, trade networks became very intricate, and after AD 1100, Indian cotton goods and Chinese porcelain were reaching very remote Indonesian islands (whose main exports were spices) and distant African ports (whence came ivory and gold).

Knowledge of dyestuffs, cotton weaving and steel-making spread through much of the region, so that despite many local variations there were close similarities in basic techniques from Iran to Indonesia. For example, methods of making sword blades of 'Damascus steel' (see chapter 5) were widely spread in Syria, Iran and India, and after 1300, comparable techniques were used in Java also.

Irrigation engineering also shows similarities over the whole region, and one may compare dams built as earth embankments in Iraq and Iran with earth dams in parts of India and in the island kingdom of Lanka (now Sri Lanka). Some of these dams are very ancient, but more construction was under way in the eleventh century both in Lanka and in the Chola kingdom of South India. One example is the reservoir known as the Giant's Tank, built in Lanka between 1157 and 1186.

In Southeast Asia, there were engineering works of comparable dimensions associated with the famous Angkor Wat temple, in what is now Cambodia. The problem here was that although rainfall was ample to support agriculture for part of the year, there was a long dry season. By building reservoirs to retain floodwater, irrigation of an extra rice crop became possible, utilizing a complex network of canals. Because the land was fairly flat, long embankments were needed to form these reservoirs, which were rather shallow. However, the Angkor state had the capacity to organize large-scale construction, and at its fullest extent, around the year 1150, some 167,000 hectares were irrigated in this way. The influence of Indian architecture is very evident in the Angkor Wat temple, and the same is likely to have been true for these associated engineering works. Moreover, trade was not the only means by which knowledge spread. Significant numbers of Indians had settled in Southeast Asia and married into leading local families.

Ships and shipbuilding must form part of our view of the technology of this region, because sea-borne trade, diplomatic travel and migration were all important. As an example of the last of these, we should note a major Indonesian population movement across the Indian Ocean to Madagascar around AD 500, which was important for introducing new crop plants into Africa, notably bananas. At Borobodur in central Java there is a great Buddhist temple complex built three centuries later whose many stone carvings include seven reliefs which illustrate ships. They include several which were certainly capable of crossing the Indian Ocean (figure 6).

Evidence from these reliefs as well as from other sources shows that one basic feature of Indian Ocean ships was that they were built by sewing planks together by means of vegetable fibres (such as rattan, from palm trees). With the planks placed edge to edge, fibres were passed through holes bored in them, which were later sealed with putty (a mixture of lime with resin or oil). The hull was built up like this as an unreinforced shell before the ribs or thwarts necessary to give it strength were inserted.

Many different kinds of vessel were constructed in this way, ranging from large Arab dhows to small boats. Ships built in India, especially those from Gujarat, often seem to have been quite large, but even so,

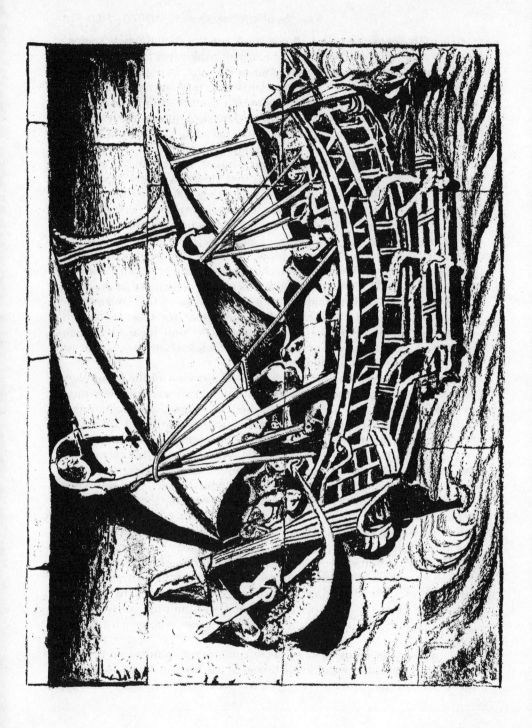

when Chinese junks (which had nailed rather than sewn hulls) visited Indian ports, they towered over local shipping, and in 1328 were described as 'like mountains with wings'.

Because of a lack of good timber in the Persian Gulf and Red Sea regions, shipyards there imported teak from South India, and Arab shipwrights also built many vessels on the Maldive Islands, which could then sail home loaded with timber and coconut fibre (coir) for rope-making. These activities provided frequent opportunities for cross-fertilization between Indian and Arab technology. Similarly, there were close contacts between India and the main Indonesian kingdoms, though the ships built in this area were of distinctive design, with outrigger floats on either side and other features reflecting their derivation from canoes. Among other distinctive features were the Indonesian tripod masts with hooked tops (see figure 6).

Travellers from India who brought ideas about architecture and irrigation to Java and Angkor may sometimes have returned to India with seeds and plants of unfamiliar crops. One result was that India played a central role in spreading South Asian plants to the rest of the world. Indian farmers became skilled at selecting crop varieties suited to their cooler climates, and many plants were grown for their medicinal use as well as for food. A further step in the spread of many useful species occurred when travellers from the Islamic world sought medicinal herbs in India, or took back ornamental plants to try in the many gardens of Baghdad. One example was a new variety of orange brought to the city in 932 and planted in a garden belonging to the caliph.

Although a number of important agricultural crops such as rice and sugar cane had already spread from India into southern Iraq, the Islamic period was characterized by a vigorous development of these crops, with new strains selected for other areas in Iran and Iraq as well as in Africa and Spain. Thus it was Islamic agriculturists who brought rice, oranges, sugar cane and cotton into Spain. It was they who introduced sugar cane into Ethiopia, and who also made the East African island of Zanzibar famous for its high quality sugar.

Figure 6 A carved stone relief showing a ship with outriggers, in the Buddhist temple at Borobodur, central Java.

The relief dates from about AD 800. The ship has tripod masts stayed with ropes, and a steering oar (shown with a human figure clinging to it). The patterns along the side of the hull are probably intended to represent its sewn construction.

(Illustration by Mildred Pacey based on photographs of the carving in Needham, *Science and Civilization in China*, volume IV, part 3, by permission of Cambridge University Press.)

The introduction of new crops was sometimes associated with the introduction of relevant tools and irrigation methods. These developments, indeed, can be seen as a technological dialogue in which Indian agriculturists and herbalists were the origin of much knowledge, with Islamic gardeners, farmers and irrigation engineers responsible for many adaptations and selections of crops and techniques to suit new environments.

Buddhist technology

Meanwhile, a quite different technological dialogue was going on between India and China, conducted under Buddhist rather than Islamic influence. Although Buddhism had entered China a long time before, it developed deeper roots from about AD 500, and about that time spread to Japan. In the seventh century especially, many pilgrims went to India and brought back texts for translation, and were also responsible for the transmission of a range of technical information. Indian monks travelled to China too, and in 664, one Indian was able to identify soils in China which contained saltpetre, and demonstrate the purple flame which occurs when this material is put into a fire. Later Chinese studies of the chemistry of saltpetre show other evidence of Indian influence, which seems to have been the starting point for the Chinese investigations which led to the first recipes for gunpowder.

Another type of stimulus, more directly related to the Buddhist religion, arose from the custom of making massive statues of the Buddha. In India and China, these were usually of stone, but in Japan, some were made of cast bronze. For example, a huge bronze Buddha was begun at Nara in 743 and completed fourteen years later. Its height was 13 metres and 380 tons of bronze were used in its construction. In order to avoid having large numbers of joints and seams in the structure, some 40–60 tons of molten metal at a time had to be poured into the mould, which was formed around an earth core. This entailed building many small furnaces around the site, each capable of melting about a ton of metal, and operating them simultaneously.

Another technology of interest to Chinese Buddhists was the construction and repair of bridges, which became a 'pious duty'.[6] This was particularly relevant to the overland routes used for travel to India. In the mountainous regions they traversed, there were precarious footbridges made of ropes strung across gorges, notably in Kashmir and western China, and it may be significant that suspension bridges made with iron chains date back to this period.

The most important invention associated with Buddhism, however, is the printed book. A Chinese monk who was in India for a time from about 671 reported that: 'The priests and laymen in India . . . impress the Buddha's image on silk or paper.'[7] This relates to the Indian technique for printing patterns on cloth from carved wood blocks, but was never developed there for printing books. The origins of that go back to the time when paper was first used as a material on which to write. This was in China, probably before AD 100. At this time, important documents were often inscribed on stone tablets to ensure their permanence. When copies were required, rather than rewrite them by hand (which could lead to inaccuracies), paper was placed on the stone and rubbed over with an ink stick. By this means an accurate copy was quickly obtained.

This technique had probably evolved into a process for printing from carved wood blocks before the contact with India mentioned above. However, the idea of carving whole pages of writing on wood and printing from these took time to develop into a regular book-production method. When it did, under the culturally brilliant Tang dynasty, one of the chief centres for printing was Sichuan, the most important area for paper-making. There are frequent references to printed documents from about the year 800, and the oldest printed book-length work which still survives – part of the Buddhist scriptures – was produced in 868. By this time, printing was known in Korea and Japan also. In the latter country, however, its association with Buddhism was so close that until as late as 1590 printing was restricted to the monasteries and to the production of religious works.

Chinese printing did not involve a press, as in the West. A suitably carved wooden block was placed face upwards on a table. It was then inked, paper was placed face down on the block, and the back of the paper was gently brushed. The technique was developed further under the Song dynasty. Movable type made of wood or ceramic material was introduced around 1050 so that it was no longer necessary to carve a whole page of writing, as it could be set up from a stock of reusable type. Many books on technical subjects were soon being printed, including works on agriculture (AD 960 and 1018), arithmetic (1074) and a range of medical topics (1026 and 1074).

Asian perspectives

By the eleventh century, then, printing was well established and printed material was already a significant means of spreading technical information within China. But printing was not adopted in India, nor in the Islamic world, and books from China did not circulate there. Whilst trade provided some contact between China and Iran, detailed technical and intellectual exchanges between China and points west rarely went further than Buddhists in India.

This meant that India was located on something of a watershed, having contacts with the Islamic world as well as China. Indian medical ideas even had influence in Europe through the writings of an Arabic writer who became known as Rhazes, and the Islamic world took up Hindu numerals and some of the ideas of Indian astronomers. One may thus feel that Asia was divided between Islamic and Buddhist areas of discourse, with India participating in both.

However, what has struck historians of technology is not so much Asia's subdivisions as its common themes, notably the fact that hydraulic engineering developed strongly in all the major civilizations, largely because of the importance of irrigation. Indeed, one can easily feel that significant differences between European and Asian technology arose from the greater stress on mechanical invention in the West as compared with the emphasis on large-scale hydraulic works in much of Asia.

Among those who have pointed to the importance of large-scale water engineering in this context, Karl Wittfogel is known for analysis of what he called 'hydraulic societies'. He stressed the very large labour force needed to construct dams and canals using hand-tools, and discussed how systems of 'corvée labour' evolved to carry out such work, with people conscripted for several weeks or months each year. In China, every family except those of officials and gentry had to provide labour when required. In some other countries, the law specified how much labour could be demanded (about forty days per year in one instance). Powerful, indeed 'despotic' governments were necessary to operate this system, and once they had such power, they tended to use it both to organize the building of temples and palaces, and also to control and tax trade (and in China, the iron industry).

It is now generally accepted that Wittfogel's arguments were too general and too deterministic. Rather than saying that powerful, bureaucratic government arose because of the need to build large engineering works, one can equally argue that large canal projects in China and big dams in Iraq were only begun because governments had first developed the administrative capability to manage them. There was nothing inevitable about the scale of construction in most cases. For example, in the rice-growing areas of South Asia, farmers have often organized construction of irrigation works on their own, or working in small groups. They rarely required massive organizations to plan, supervise or provide a labour force.

But if Wittfogel's ideas about Asian 'hydraulic societies' are no longer really convincing, his theory is a useful reminder of the close relationship between technology and institutions. The kinds of technology which any society develops must depend on its ability to mobilize labour, to foster relevant skills, and to encourage innovation. Such capabilities depend in turn on the effectiveness of commercial, industrial, and governmental institutions.

There will not be space in later chapters to give these topics the emphasis they deserve, but we shall certainly see sharp contrasts between parts of the world where military institutions were a dominant influence on technology, and regions where agrarian bureaucracy or commercial institutions were more important. It will be tempting to think that the most creative societies were those in which *many* types of institutions were active and in dialogue with one another. Europe may later have had that quality, but for the moment, our example is eleventh-century China. Here there were governmental institutions, independent entrepreneurs and Buddhist monasteries all playing a part, notably in metalworking techniques and in the development of printing.

A corollary of this is that when institutions declined, or were disrupted by a conquering power, there was often a *loss* of technology. The best-known example from before the period discussed here was the collapse of the western Roman Empire in the fifth century, when engineering skills disappeared along with the institutions which had organized their use. That is one reason why the West was relatively backward during the period covered by this chapter. Other examples are the decline of Baghdad from about 1050, and of North India in the 1190s, as the next chapter will show. To some extent, indeed, the story of Asian technology in later centuries is a history of setback and loss during periods of institutional disruption or decay. The chief examples are China from the middle thirteenth century and again in the nineteenth, and India throughout the nineteenth century. By the end of that period, Japan was conspicuous as one nation in which a rapid development of both new and existing institutions had enabled increasingly elaborate technologies to be handled effectively.

2   Before the Mongols

Asian conflicts and contrasts

One of the major themes of Asian history is the relationship between the nomadic peoples of the northern grasslands and the settled civilizations of China, India and the Iranian plateau. Although the nomads had no engineering skills to speak of, and few handicrafts, their significance for the history of technology is considerable. This has already been illustrated for China, where efforts to control nomadic encroachments from the north led to a massive expansion in the production of weapons and the use of iron. More important, nomadic invasions increasingly led to conquest: from the 1030s in Iran, from 1126 in northern China, and from the 1190s in India. Then came the most formidable nomadic power of all, that of the Mongols. Their invasions of both Iran and China extended into the 1260s, and completed two full centuries of upheaval. This can be taken as ending the most creative phase of the 'age of Asian technology' whose culmination was described in the previous chapter.

The nomads gained their livelihood by raising sheep and horses. They had little understanding of the irrigated farming systems of the lands they conquered. As they took over government in Iran and Iraq, and later in North India, irrigation canals were sometimes wantonly damaged, and their maintenance was neglected. Agriculture declined, most notably in Iraq, where clogged drainage channels contributed to a build-up of salt in the soil. As food production fell, the population of Baghdad declined quite steeply.

That is one reason why we may see the nomadic invasions as marking the end of an era in which irrigation engineering was central to many other technologies. Another factor was the disruption of centres of expertise, and the destruction of libraries, notably at Baghdad and in India. A more long-lasting result was that, after the invasions, ruling groups in the Islamic world, and in China also, became more conservative in outlook and ideology and with respect to technology.

With the decline of Baghdad from about 1050 onwards, some scholars moved west and Spain became the greatest centre of Islamic learning and technical knowledge. Thus the two centuries of disturbance which ended the earlier innovative phase in Asian technology also saw the transfer of some Asian expertise to a more westerly location. This will concern us greatly in the next chapter. The purpose of the present chapter, however, is to review the diversity of Asian technology just before and during the period of upheaval. The point here is that we have so far examined common themes and widely used techniques in Asia, such as hydraulic engineering. But the utterly different lifestyle of the nomadic warriors who are now part of the picture forces us to notice some significant contrasts.

The people who invaded Iran and later India in the eleventh and twelfth centuries mostly spoke Turkish languages, and some had adopted the Islamic faith. After a long period of encroachment in northern India, a full-scale invasion was mounted in 1192. Large areas around Delhi and extending into Bengal were conquered within a fairly short time and a new Islamic state was set up, known as the Sultanate of Delhi.

It was only after this, in 1206, that the Mongol tribes decided to unite under a single leader. Then, the man at their head took the title of Chingiz (or Ghengis) Khan. Given that almost every adult male was an expert horseman, skilled in archery and practised in hunting, the result was the creation of a formidable military force whose impact on China, Iran and Europe will be mentioned again in the next chapter.

For the purposes of conquest, if for nothing else, Turkish and Mongol nomads clearly had a very effective technology. However, their lifestyle was basically simple, centred almost wholly on their flocks and herds. Sheep were a basic source of food, and supplied both meat and milk. Their skins were used for clothing, and their wool was woven on a loom to make a felt-like fabric for making the tents in which people lived. The looms on which this weaving was done had to be portable and so had no frame. Thus a loom consisted of beams and rollers which were held by pegs driven into the ground. A tripod was used to support other components (notably the heddles).

The nomads were not quite self-sufficient, however. They sold horses into China and India so that they could import grain, tea, textiles and especially iron for making swords, arrow-heads and stirrups. Horses were central to the nomadic economy, therefore, partly as an export and means of transport, partly as a source of milk (which was fermented to make an alcoholic drink), but most important of all, for riding when rounding up sheep and for hunting (often a vital source of food).

The key technology associated with the nomads' horses was the harness, which allowed the many functions associated with these animals to be performed. The crucial element here was the iron stirrup which, in conjunction with a suitable saddle, enabled a rider to hold himself steady whilst using both hands to shoot arrows from a bow, either in hunting or in warfare (figure 7). The stirrup had evolved in China and Mongolia five centuries before, but was still not universally used. This gave the nomadic armies with their mounted archers a decisive advantage in many conflicts. During the Turkish invasion of India, the Hindu defenders had fewer horses and used a more primitive wooden stirrup or possibly no stirrup at all.[1]

The bow used by Turkish and Mongol archers was a compound device which used animal products such as horn and sinew as well as having a wooden frame. It was stiffer to pull than the English longbow, even though it was fired from horseback. It had considerable range and a lethal effect.

After Baghdad fell to the Mongols, Delhi became, for a time, the most important Islamic capital east of the Mediterranean. Scholars and scientists took refuge there, though having lost their libraries they could not entirely recreate the brilliance of Baghdad as a seat of learning. Even so, Greek mathematics was taught in Delhi, and according to Irfan Habib, the eminent historian of Indian technology, new techniques spread into the region also. For example, he suggests that the magnetic compass was first used on Indian ships at this time, and that new centres for paper-making developed. Indian documents written on paper survive from before this time, but the number is much greater from the thirteenth century onwards.

Figure 7 Nomad warrior on horseback, with characteristic saddle, stirrups and bow. Note that stirrups are essential for a mounted archer.
  (Illustration by Hazel Cotterell)

Whilst Islamic domination of North India may have had these positive aspects, the initial conquest by Turkish-speaking armies in the 1190s did great damage to Indian learning, both technical and general. The conquest was particularly destructive in Bihar and Bengal, where Buddhist monasteries were sacked and many monks were killed. One consequence was the virtual elimination of Buddhism in the region, which is where it had originated seventeen centuries earlier. In 1194, the great centre of Indian learning at Benares was attacked, and numerous monuments as well as books, records and probably an astronomical observatory were destroyed. The scale of this vandalism was probably a lasting setback for Indian science, and astronomy was not again seriously studied until the fifteenth century. The last celebrated Indian astronomer for a long time was Bhaskara, who was working in the 1150s, and whose writing had some influence in the West.

Spinning wheels and winding machines

One controversial aspect of the history of technology in North India[2] concerns the spinning wheel. Because cotton textiles originated in India, it has long been assumed that the spinning wheel, as used with cotton, must also have been invented in the subcontinent, perhaps between AD 500 and 1000. However, early references to cotton spinning are so vague that none clearly identifies a wheel, according to Irfan Habib. The references could equally indicate earlier methods of hand spinning. The earliest unambiguous reference is in a document dating from about 1350 which mentions women using spinning wheels in the previous century. Habib also points out that the most usual word in India for a spinning wheel is *charkha*, and this derives from the Persian language. He therefore thinks that the spinning wheel was introduced into India from Iran in the thirteenth century.

If the spinning wheel is not an Indian invention, where did it originate? The earliest clear illustrations of this machine come from Baghdad (drawn in 1237), China (*c*.1270) and Europe (*c*.1280). It might seem from this that spinning wheels appeared in China and Europe at almost the same time, and one may be tempted to think that the conquests of Mongol armies both in Eastern Europe and in China had created conditions for especially rapid transmission of new techniques. This may be true of some inventions, as the next chapter will show, but is not convincing in this case because simpler forms of wheel for winding bobbins were in use much earlier, and some silk-processing equipment was also widely disseminated before the Mongol conquests began. Alongside these machines, there is some evidence that spinning wheels of some sort may have already come into use in both China and the Islamic world during the eleventh century.

However, in view of the vagueness of the evidence, arguments about origins should not be pushed too far. It may be useful to think again in terms of a technological dialogue in which the westward dissemination of silk and cotton textiles from China and India respectively stimulated many local responses. These may have included numerous minor innovations, and once the winding wheel was known, some form of wheel for spinning may have been suggested to the minds of a number of individuals in quite different places.

The possibility of independent regional invention was all the greater because of the distinctive qualities of textiles manufactured in different places. For example, surviving specimens of silk fabric show that thread produced in the Islamic countries (and Europe) was always twisted, to make it stronger, whereas Chinese manufacturers seem to have avoided this because fabric woven from twisted thread was stiffer and less lustrous. For that reason, mechanical devices for twisting thread are more likely to have been invented in the West rather than China. Indeed, there is an Islamic description of a silk twisting machine dating from just before 1030. Experience of this machine could have led to a distinctive approach to spinning other textile fibres, and may ultimately have contributed to the invention of the specifically European type of spinning wheel known as the Saxony wheel.

Some emphasis on silk manufactures is justified because of the suggestion by some historians[3] that, in China, mechanized techniques developed first in silk workshops and were then adapted for use with cotton and other fibres. In this context, the first process to consider is unwinding the long, continuous filament from the silk-worm cocoon. This was done by plunging the cocoons into hot water, which killed the 'worm' whilst loosening the filament. At quite an early date, and certainly before 1090, silk-reeling machines were being used in China which had a small heated bath of water containing the cocoons at one end and a large reel onto which the filaments were wound at the other (figure 8).

Two other processes were necessary before silk thread or yarn was ready for weaving. Firstly, several filaments had to be combined in a single, strong thread. This is where Islamic workers chose to twist the filaments together. In China, thread was often produced without twisting, by relying on the natural gum from the cocoon which coated each filament. This could hold the filaments together once they had been brought into contact on a winding wheel adapted for 'doubling' threads. The gum protected and strengthened the thread during weaving, but was eventually removed by steeping the finished cloth in water.

The final process before weaving was to wind the thread on to bobbins

Figure 8 A Chinese silk-reeling machine, drawn in 1843, but corresponding closely in all mechanical detail with a description of this type of machine written about the year 1090.

The operator is unwinding silk-worm cocoons in a heated bowl of water while slowly rotating the reel via a treadle and crank mechanism (shown in rather odd perspective). The two silk filaments being drawn from the cocoons are led over rollers at the top of the machine, and then through guides (A) and onto the reel. An endless belt on the axle of the reel is turning a pulley (B) on a vertical axis. The crank attached to this pulley moves the rods above it and thereby moves the guides slowly a few centimetres to and fro. This ensures that the silk is laid down in broad, even bands on the reel.

(From Needham, *Science and Civilization in China*, volume IV, part 2, figure 409 and p. 107, by permission of Cambridge University Press.)

of convenient size for use at the loom. It was for this operation that textile workers invented the bobbin-winding wheel, perhaps at a very early date. This was referred to earlier as antecedent to the spinning wheel. One striking development after this was that the Chinese built machines working on the same principle as the winding wheel, but capable of rotating ten or more bobbins simultaneously. Finally, and some time before 1300, they produced the mechanism powered by a water-wheel which is illustrated in figures 9 and 10. This is closely similar to the ten-bobbin winding machine, but in 1313 it was referred to as the 'great water-driven spinning machine', for use with hemp or ramie, two vegetable fibres of some importance in China.

The surviving illustrations of this machine present a number of problems of interpretation. As they appear in existing copies of a book on agriculture by Wang Zhen, they are wood-block prints, the lines on which seem to have been copied by somebody who did not understand what was being represented (figure 9). The bobbins at the bottom of the machine were rotated by an endless cord and as they turned, thread was wound on to them from the large reel above. But the thread would pass through guides of some sort between the reel and the bobbin, perhaps like the guides on the reeling machine (see figure 8). Moreover, if this machine was really spinning yarn from a fibre such as ramie, there would have to be a mechanism to insert some twist into the thread. Such details are not shown in Wang Zhen's illustration, and are omitted from the reconstruction presented in figure 10, which also uses evidence from photographs of similar machines as used in China in the nineteenth century. Thus figure 10 shows a winding machine, not a true spinning machine. Wang Zhen had a separate drawing of the water-wheel and its rope-and-pulley drive, so this part of the reconstruction is unambiguous.

Although there were Chinese machines which depended on wooden gear wheels, the Chinese preference for transmission of mechanical power via ropes or belts and pulleys should be noticed.

Survival technology

Mechanical techniques reached a high level of sophistication within the Islamic civilization as well as in China, notably with respect to applications of water-wheels. India presents a different picture, however. We may have a misleading impression because of the vagueness of most available information, which may be due to the destruction of records and equipment during the 'Turkish' conquest of the 1190s. However, it would seem that in India and southern Asia relatively little use was made of machines. There were certainly some water-raising wheels, and rice

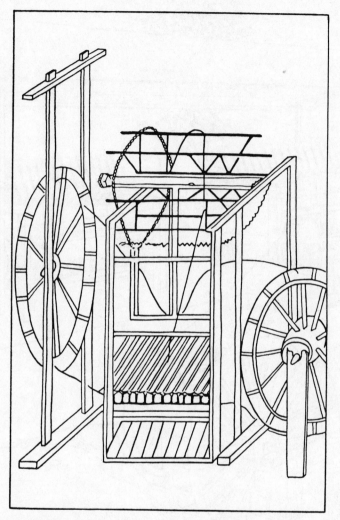

Figure 9 Illustration of a 'great water-driven spinning machine' from Wang Zhen's *Agricultural Treatise*, first published in 1313, reprinted *c.*1530. A separate illustration showed the water-wheel which drove the machine.

The diagram was probably redrawn for the 1530 edition, and is not entirely clear.

(From Needham, *Science and Civilization in China*, volume IV, part 2, figure 627a, by permission of Cambridge University Press.)

farmers in Java and Bali at some unknown date developed their own mechanical device for lifting water. It consisted of a pivoted bamboo tube or trough which would fill with water from a stream, then tip over under the weight of the water, lifting some of it to the level of the field. But even in India, there are no reports of machines with gears, pulleys, cams or cranks before 1200 – that is, if it is correct that neither the spinning wheel nor the geared type of water-raising wheel (the 'Persian wheel') was introduced until after that date. Yet it cannot be said that India was lacking in technology, for the subcontinent's contribution in terms of food crops, textiles, medicine, minerals and metals was clearly very important, as the previous chapter showed. Lack of machines did not mean 'backward' technology. It indicated, rather, a different emphasis as compared with the kinds of technological development characteristic of Iran and China.

One way of analysing differences of this sort is to notice Francesca Bray's comparison of agricultural methods in two regions of China.[4] In the north, the cultivation of wheat had developed by AD 1000 to the point where animal-drawn ploughs, harrows, seed-drills and hoes were operating in large fields. The ploughs had iron shares and mould-boards, and the harrows had iron tines. Compared with this, the rice culture of South China could seem less advanced, since fields were small and hand-tools were often used. However, Bray points out that a very sophisticated complex of complementary tools was available, including specialized hand-tools for weeding, as well as winnowing machines and mills for processing the crops after harvest. Thus South China was not 'backward' compared with the north, but had adapted just as cleverly to a very different crop and a different environment.

Bray characterizes this situation by saying that each area had a distinctive *tool complex*, and, taking a slightly larger view, it is possible to speak about the *technology complex* of a particular region.

One way of beginning the analysis of technology complexes is by looking at the techniques used by different peoples to obtain their food and other basic necessities. For example, it is remarkable how completely the nomadic peoples mentioned earlier depended on animals for a variety of foods, and also for clothing and shelter. By contrast, the peoples of South India and Southeast Asia were almost equally versatile in the use

Figure 10   Reconstruction of the 'great water-driven spinning machine' shown in figure 9, together with the water-wheel which drove it. Wang Zhen's book gives details of the water-wheel, but not of how many machines it drove, so that part of the drawing is conjectural.
(Illustration by Hazel Cotterell)

of crops and trees to produce food, fuel, medicine, clothing (from cotton) and housing (of wood with palm-leaf thatch). Sometimes farming with trees and crops was closely integrated, especially in rain-forest areas where retention of trees helped protect the soil from erosion by heavy rain. In parts of Java, for example, not only were there wet fields for rice, but trees were planted on land that was also used for vegetables or sugar cane so that the land would yield a harvest at two levels – from the vegetable crop at ground level and from the fruit or spices on the trees above. The term 'multistorey farming' is apt, and in South Asia systems of this sort were developed in many ways. Sometimes open areas for wet rice cultivation were bordered with trees. Sometimes the trees which produced cloves, the major export crop of the Indonesian islands, were incorporated. By contrast, some communities on the smaller islands depended heavily on fishing, and so boats were crucial in their basic technology complex.

Thus, comparing nomadic horsemen and pastoralists with South Asian farmers, and contrasting South China's rice culture with wheat farming in the north, we have quickly identified four different technology complexes related to production of food and other necessities (table 1).

However, there are not only contrasts between people who kept animals and those who kept different kinds of crop. There is also a contrast in the scale of operations, and in the amount of engineering involved.

In communities practising rice culture or other productive forms of farming, individual farms could produce a lot of food even if they were very small. Moreover, on small farms where there was a rainy climate, the construction work necessary to provide irrigation for rice needed only a small labour force. In North China, by contrast, farms were bigger, and there was more advantage in using animal-drawn implements. Engineering works were sometimes on a massive scale, both to provide irrigation in dry areas, but more especially to control flooding from the mighty Yellow River. In Iran and Iraq there was a comparable technology complex (see table 1).

It is hardly surprising, then, that there was a bigger engineering content in the technology complexes of North China and Iran (figure 11) than in that of South Asia, where large-scale works such as those at Angkor were in some respects an anomaly. This was an inevitable reflection of differences in environmental conditions. But apart from conditions relating to climate, soils and ecology, two other factors ought to be considered.

Firstly, institutions relating to land ownership were clearly important. Large farms or big estates (North China) and the exercise of ownership rights by kings (Angkor) all contributed to the large scale of engineering

Table 1   Five Asian technology complexes

Region	Basic 'survival technology'	Machines	Engineering works
IRAN AND IRAQ	Irrigated agriculture	*Extensive use* (water-wheels, windmills, spinning wheel, gears, cams, pulleys)	*Large-scale* (dams, canals)
NORTH CHINA	Dryland and irrigated agriculture	*Extensive use* (water-wheels, spinning wheel, gears, cams, pulleys, cranks)	*Large-scale* (canals, flood-control works)
SOUTH CHINA	Wet rice culture	*Less extensive* than North China, but similar	*Mainly small-scale* (small reservoirs or ponds for rice irrigation)
SOUTH INDIA and SOUTHEAST ASIA	Wet rice culture and tree crops	*Very limited* (water-raising devices)	*Mainly small-scale* with exceptions in Lanka and Angkor (small reservoirs or 'tanks' for rice irrigation)
CENTRAL ASIAN GRASSLANDS	Animal husbandry	*Almost none* apart from portable looms	*None*

in some places. Secondly, population density was always crucial. When large numbers of people had to be supported by a small area of land, more elaborate irrigation works might be required than for a lower population in that area, but at the same time there would be more labour to construct such works.

These points refer chiefly to technology for the production of food and basic necessities – that is, 'survival technology'. Other aspects of a regional technology complex depended on institutions such as the Buddhist monasteries in China and Japan, whose role with regard to the development of printing and certain kinds of metalwork was mentioned in the previous chapter. Institutions connected with government, markets and the military were often of even greater importance.

This should make it clear why it is so often inadequate to think of 'transfer of technology' from one culture to another without recognizing the modifications in technique that are likely to follow, and the entirely

fresh inventions which may emerge in response. Where survival technologies are concerned, environmental conditions are likely to render many alien techniques irrelevant and require major modifications of others. The distinctive requirements of local institutions may have a similar influence.

Fine technology

When looking in detail at the institutions which influenced technological development, one should note the patronage sometimes extended by the wealthy or powerful to individual craftworkers, and one should take account of the role of observatories, such as the ones at Baghdad and Kaifeng, in providing scientific and technical advice to governments.

Observatories could have practical importance in map-making, or for checking that a reliable calendar was available. The latter was important if governments were to levy taxes and celebrate religious festivals on the correct dates. However, the role of observatories was not always so practical.

Technology was always to some extent involved in the dreams and fantasies of a society or its rulers. Often these were fantasies of dominance and power, such as the Mongol dream of conquering the whole world, initiated by Chingiz Khan. But some fantasies were related to political theory, such as the idea that the authority of the Chinese Emperor derived from a 'mandate of heaven'. That was another reason why he had an observatory, because its astronomers were presumed to understand the heavens.

In the Islamic world, there were also gentle dreams about which poets wrote, and these stimulated invention, particularly in a field which the engineer and historian Donald Hill describes as 'fine technology'. This

Figure 11   Large masonry dams diverting water into irrigation canals were a significant aspect of 'survival technology' in Iran and Iraq, as were water-powered corn mills like the one being constructed here.

The dam is typical of several built during the eighth to eleventh centuries, and is surmounted by a bridge whose pointed arches are very characteristic of Islamic architecture. The small mill is powered by a horizontal water-wheel more typical of mountainous areas with small streams and waterfalls, but also known to have been used at dams in Iran. Water to drive the wheel would flow from behind the dam via the large pipe (made of wood) on the left of the picture.

(Illustration by Hazel Cotterell based on information given by Norman Smith, al-Hassan and Hill, also Wulff.)

term denotes small but relatively sophisticated mechanisms of which we may note two kinds, one connected with gardens, the other with astronomy.

With regard to the first category, it is significant that the garden poem became a distinct genre in Islamic literature, drawing on earlier Persian traditions. For people living in an environment dominated by hot, dry deserts, gardens with fruit trees, shade and running water were understandably attractive, and often featured in serious poetry and in more popular literature such as *The Thousand and One Nights*, compiled around AD 850.[5] It is typical that references to gardens and quiet courtyards often emphasized the fountains they contained, and that the technical literature of the period also includes unusual stress on fountains and how they worked.

An early example in a book by the Banu Musa brothers, also written around the year 850. The brothers exemplify the role of institutions associated with state patronage, for they were members of an academy founded by a caliph of Baghdad and connected with an observatory. One of several fountains they described had a jet of water turning a small horizontal wheel by playing on turbine-like blades from below.

Very much later, in 1206, an Islamic technician named al-Jazari wrote a book on 'mechanical devices' under the patronage of a local ruler in northern Iraq. Figure 12 represents an interpretation of a design for a fountain by al-Jazari. It could work with either a single, vertical jet of water, or a ring of six jets rising from a spherical container. This required separate supplies of water, which came in separate pipes, one inside the other.

The other aspect of fine technology we need to notice was concerned with making astronomical instruments and computing devices. The central concept underlying this work was that the motions of the planets were governed by a complex system of circles. This was a mathematical concept used for calculation, and it stemmed from a book written by Claudius Ptolemy, the last great astronomical writer in the Greek tradition, who lived from about AD 85 to 165. But Ptolemy's mathematical circles could be visualized as great wheels, one revolving in another, and were so described in the famous series of poems known as *The Ruba'iyat of Omar Khayyam*.[6]

Writing in northern Iran about the year 1100, Omar Khayyam was an astronomer and mathematician. He made frequent references in his poems to the 'wheel of the heavens', once imagining it modelled in a darkened room with 'the sun the candle' and 'images revolving on the walls'. That is a significant metaphor, because many kinds of mechanical device were being made at this time which were conceived in such terms. One was a clock mounted above a gateway in the city wall at Damascus.

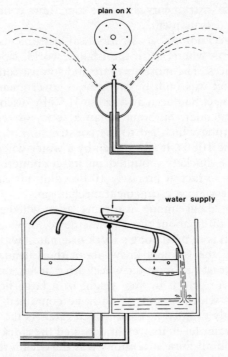

Figure 12   Fountain designed by al-Jazari with (top) provision for a single central jet surrounded by six smaller jets, and (bottom) the cistern supplying the fountain. The pipes and tanks were probably made of sheet copper with soldered joints.

The water supply was fed into the funnel in the centre of the cistern. This was attached to a pipe which could swing over to supply first one tank and then the other, so that different jets on the fountain would play alternately. In the position shown, the right-hand tank is filling and a tipping bucket will eventually tilt over and thereby push the pipe over to fill the other tank.

(Illustration from Donald R. Hill, *A History of Engineering in Classical and Medieval Times*, by permission of Croom Helm, Publisher.)

It represented the time by displaying a model sun moving across the sky. This was not the only clock made in the twelfth century to represent the sun so directly. Others had candles or lamps at night to represent heavenly bodies, as in the poem.

These clocks were something of a spectacle, with mechanical animals, birds and 'jacks' performing every hour. By contrast, many very precise instruments were made for serious astronomical observation, for use in astrology, and for navigation during journeys across the desert as well as at sea. But these instruments also were sometimes conceived as models of the 'wheel of the firmament'.

A similar 'fine technology' of astronomical instruments and clockwork developed in China where, as in the Islamic world, clocks were driven by hydraulic devices. The most accurate and sophisticated astronomical clock of the period was built by Su Song, a government official whose interest in the subject had been aroused in 1077 by discovery of an error in the Chinese calendar. Su Song began a series of researches on the measurement of time which led to the construction of a large clock at Kaifeng during the 1080s. It was driven by a water-wheel, 3.4 metres in diameter, with 36 buckets mounted on its circumference. The wheel revolved slowly – in fact at precisely 100 revolutions each day – under the control of a very exact escapement mechanism.

There is disagreement among historians[7] as to whether Islamic clock-makers knew anything about this Chinese device or vice versa. The one certain connection is that Su Song's work originated with concern about the calendar, and the Chinese knew about the calendar used in Iran. However, in contrast to the water-wheel drive in Su Song's clock, most Islamic clocks were driven by the weight of a large float – perhaps a block of wood – which would typically be contained in a tank from which water slowly drained at a controlled rate. As the float descended with the falling water level, its weight operated the clock mechanism.

One idea behind all these inventions was the dream that if one could make a clock or other instrument that exactly reproduced the motions of the sun and the planets, one would capture something of their essence. At a time when many people believed in astrological forces emanating from the planets, this was a powerful concept. But as Omar Khayyam commented with characteristic scepticism, 'Nobody has mastered the wheel of the firmament.'

Ptolemy's astronomical theories had also been studied in India for a very long time, one Indian book on the subject dating from about AD 500, and we may deduce a certain amount about Indian mechanical technology from the way this subject was tackled. Around 1150, the astronomer Bhaskara (who has already been mentioned) wrote a book in which there was a chapter on astronomical instruments. He had been observing water-raising wheels of the type shown in figure 4, and seems to have imagined such a wheel lifting enough water to replenish the stream driving it. If that could be achieved, the wheel would turn perpetually without any extra water entering the stream. This was another dream which was to beguile inventors for centuries, and Bhaskara's version of it, in which the water-raising wheel was replaced by a hollow wheel 'like a *noria*' containing mercury, was to have wide influence.

The years of upheaval

Many of the developments described in this chapter were interrupted or even terminated by nomadic invasions. Iran had been conquered by Saljuq Turks just before Omar Khayyam's birth, and he grew up with evidence of economic decline all around him. The pessimistic tone of his poetry may well be a reflection of this.

In China, Su Song's clock was dismantled in 1126, when invaders conquered Kaifeng (though the clock's parts were kept, and survived until the advent of the Ming dynasty). Two centuries later, after the more devastating Mongol conquest, Wang Zhen mentioned the destruction which had occurred as one reason for writing his book on agriculture (in which his account of the 'great spinning machine' appeared). He hoped that the book would be a contribution to the rebuilding of the economy. He might well have mentioned the decline of the iron industry also. By 1260, production in the Hebei area had shrunk to only a quarter of peak eleventh-century output, and the iron was being used exclusively by the Mongols to make weapons.

Another contribution to economic reconstruction during Wang Zhen's time was made by a woman named Huang Dao-po, who travelled the countryside teaching people about 'improved implements for spinning and weaving', and possibly about cotton gins also. Her work was so much appreciated that a memorial to her was erected in 1337. A story which complements this concerns a road built through mountains in Fujian province. It was constructed in 1315 with a woman as its engineer.[8] One wonders whether this was an emergency measure, compensating for a loss of skilled men in a very disturbed period. However, given the evidence quoted earlier about the inventiveness of Chinese society stemming from many diverse sectors of society (including Buddhist monks and government officials), it seems possible that women also contributed in some areas of technology. This is especially likely with regard to textile machines, since women were the chief users of spinning wheels and worked other textile processes. Be that as it may, these stories belong to a time when institutions were changing radically under the impact of Mongol government, and when the economy was recovering with difficulty after a period of devastating warfare. The wider consequences of these disturbances will be discussed in the next chapter.

3 Movements West, 1150–1490

Islam and Africa

Paper, the magnetic compass and a new type of loom were three innovations which appeared in western Europe soon after 1150. A few years earlier, the first cotton cloth to be woven in West Africa was produced. These developments indicate that new areas were being drawn into the technological dialogue described in previous chapters, due especially to events in Spain and nearby areas of Africa.

Europe, of course, had already developed a vigorous, technically innovative culture whose achievements will be discussed later in this chapter. South of the Sahara, there had been much more limited but still significant developments in agriculture and metalworking. This region had a small and scattered population whose main centres were cut off from one another by vast areas of desert and impenetrable forest, so the diffusion of technology was slow.

The Islamic civilization in North Africa had its European bridgeheads in Spain, and until the 1070s in Sicily, and had also built up trade across the Sahara Desert with areas where gold could be obtained. In these parts of Africa, there was a fairly varied agriculture, with sorghum and millet the main cereal crops in the drier areas, and a local species of rice in districts with sufficient water.

By contrast, the densely forested areas near the West African coast were sparsely inhabited, and there were few crops which could be grown in the very humid climate until bananas (mentioned in chapter 1) and Asian yams were introduced after AD 500.

Migration of people with relevant skills has been suggested as an explanation for the diffusion of metalworking and mining techniques. The northeast of Africa, having contacts with Arabia in pre-Islamic times, is thought to have been the source of many techniques. At first people from this area moved mainly westwards and contributed to the expansion of gold exports across the Sahara. Then there were southward migrations exploiting the Katanga copper belt and the gold, copper, tin and iron of the Zambezi River area. Powerful and extensive kingdoms developed wherever there was an opportunity for dominant groups to control the trade in metals, with rulers often symbolically identified as skilled metal smiths. One such kingdom roughly coincided in its territory with what is now Zimbabwe, whereas in West Africa others of importance were named Ghana (to the north of modern Ghana) and Kanem (in the region around Lake Chad).

The chronology of these developments is vague until such time as regular contacts with the Islamic centers of North (or East) Africa were established. In 1076, Ghana was invaded by armies of the Islamic Almoravid (or Murabit) dynasty which ruled Morocco and later took over southern Spain. One result was an influx of desert nomads with large flocks and herds, destroying the fields where the Ghanaians had cultivated their millet. Because of this destruction, when a new kingdom named Mali was established, encompassing Ghana's former territory, its centre was further south. From this time on, the Negro families who ruled Mali and Kanem adhered to the Islamic faith, partly because this gave them status with the merchants from North Africa with whom they traded. Some Islamic schools were set up, and a small educated class emerged which could carry on the correspondence and accounting necessary in trade.

West Africans now sometimes went on the pilgrimage to Mecca, travelling via Cairo, and saw ploughs, wheeled vehicles and other aspects of Islamic technology. But the plough did not advance further into Africa than the Nile valley and Ethiopia, and the camel was too efficient for transport in desert and semi-desert areas for wheeled transport to be attractive. Innovations were made in West African agriculture, however. New crops introduced included citrus fruit and the Asian species of rice. One king of Kanem organized experiments with growing sugar cane.

The adoption of Islam meant new ideas about architecture as mosques were built, and a social revolution in dress. The Islamic custom was that people of both sexes should be clothed from neck to ankle. In some areas it was only the rich who adopted Islam, and poorer people continued to go nearly naked. Al-Idrisi, who wrote a book about West Africa and Spain around the year 1100, noted that in the region of the Senegal River, 'the rich wear clothes of cotton, the common people dress in wool.'[1] This was because cotton cloth was still an expensive import. However, the growing of cotton, with spinning and weaving techniques, was being introduced at this time. Documentation of the equipment used is non-existent, but deductions can be made from a distribution map of different types of loom in Africa prepared by a specialist on ethnic textile traditions[2] (figure 13). If we take only the looms used for weaving cotton, and exclude types thought to derive from later European influence, the distribution coincides almost exactly with the areas which were under Islamic influence by 1150 or soon after. Moreover, the vertical cotton loom used in the Mali region of West Africa (and operated only by women) was of a type also found in North Africa (see figure 13). It is possible that looms of this type were in use as early as 1150, and that they were introduced into Mali about then as a result of trade with North Africa. In all the Islamic areas, whilst some cloth was imported, cotton seems to have been grown and was spun and woven locally.

Islam and Europe

In Europe there were also innovations in weaving. A new type of

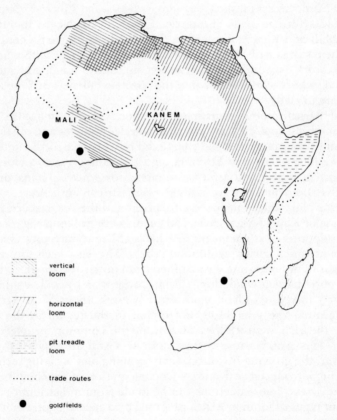

Figure 13   Distribution of cotton-weaving looms of pre-1500 type, and gold-mining areas in Africa.
   (Modified from H. Ling Roth, *Studies in Primitive Looms*, by permission of Calderdale Museums, Halifax.)

horizontal loom was introduced, notably in the Low Countries for weaving woollens. Its great advantage over earlier European looms was that some operations (such as raising and lowering heddles) could now be controlled by foot pedals, leaving the weaver's hands free to pass the shuttle backwards and forwards. The idea of pedal operation may have come into Europe from Islamic weaving. However, in Iran, Syria and parts of East Africa, when pedals were used, the operator (usually a man) sat with his feet in a pit below a fairly low-slung loom. In the West, the whole mechanism was raised higher above the ground on a more substantial frame. Looms of this type were very widely used in the Islamic part of Spain by 1177, and it was probably from here that they were first adopted in Christian Europe.

Paper-making also came to Europe via Spain at this time. The paper was made from the same vegetable fibres as linen cloth (and usually from linen rags), which first had to be pounded in water until a pulp was formed. The process had been invented in China long before, where it replaced an even older method of making paper-like material from mulberry bark. Knowledge of the technique entered the Islamic world in AD 751 after a battle in Central Asia between Chinese forces and an Arab-led army. Chinese prisoners-of-war skilled in paper-making set up a workshop in Samarqand, and from there other workmen went to Baghdad. However, paper made by the Chinese method for scribes to write on with brushes was not so good for people who used pens. Thus paper-makers supplying the Baghdad market began sizing their product with starch to achieve a parchment-like surface.

The manufacture of paper meant that books became more widely available. By AD 900 there were over a hundred shops in Baghdad employing scribes and binders to produce books for sale, and soon there were even some public libraries.[3] With growing sales of books elsewhere, paper-making spread west, as table 2 shows. Eventually, paper was being manufactured at Fez, in the Almoravid kingdom of Morocco, where the Islamic university presumably stimulated demand.

In 1085, the Almoravids were called on to move troops across the straits of Gibraltar and into Spain. The Islamic community there was in crisis, having lost the city of Toledo to a Christian army which now threatened to move further south. The arrival of forces from Morocco halted this advance and inaugurated a century of relative stability. Spain attracted Islamic scholars from more troubled regions, and Toledo became a centre of learning where Islamic and Jewish scholars worked with Christian Europeans. One practical consequence was an increasing requirement for paper, and it is no coincidence that a paper mill was operating in southern Spain by 1151.

When paper-making spread from here into other parts of Europe, the

Table 2   The diffusion of paper-making
The dates refer to paper as used for writing. Paper-like fabrics, some made from mulberry bark, were also used in clothing and as wrapping material in China, Southeast Asia, and the Pacific islands, and in this form paper may have originated as early as 200 BC in China.

Place	Date when manufacture of paper began	First report of water-wheels driving the pulping process	Comment
CHINA	AD 100		
TIBET	AD 650		
INDIA			
Buddhists	AD 670		Possibly paper imported from Tibet
Delhi Sultanate	After 1258		
Bengal	1406		
CENTRAL ASIA			
Samarqand	AD 751	AD1041	Chinese workmen in 751
OTHER ISLAMIC COUNTRIES			
Baghdad	AD 794	AD c.950	Probably Chinese workmen at first
Cairo	AD 850	–	
Damascus	c.1000	c.1000	Existence of paper mills known, but dates uncertain
Tripoli	c.1000	–	
Sicily	c.1000	–	
Fez (Morocco)	c.1050	–	
Jativa (near Valencia, Spain)	1151	1151	Jativa was under Christian rule from c.1238
EUROPE			
Spain⎱ Sicily⎰			See under Islamic countries
Italy, Fabriana	1276	1276	
France, near Ambert	1326	1326	
Germany, Nuremberg	1390	1390	
England	c.1490	c.1490	

Sources: compiled from data quoted by Tsien Tsuen-Hsuin, al-Hassan and Hill, Liu Guojun and Zheng Rusi, also Jean Gimpel.

laborious process of pounding fibres into pulp was always carried out mechanically, with a water-wheel driving a cam-shaft to operate trip hammers. It used to be thought that this was a distinctively European development. However, it is now known that there were numerous water mills in the vicinity of Baghdad, and that water power was applied to paper-making in that region two or more centuries earlier than in Europe (see table 2).

The fall of Toledo in 1085 was of great significance for Christian Europe because this was an important centre of learning where Europeans gained access to Islamic technical books, to information about Indian medicine and Hindu numerals, and to Arabic versions of Greek mathematical works. The men who did most to make this great store of knowledge available in Europe were probably Adelard of Bath (who worked mainly on Euclid's geometry and Islamic trigonometry) and also Gerard of Cremona. Living at Toledo from about 1150 until his death in 1187, Gerard seems to have organized a regular team of Jewish interpreters and Latin scribes, through whose efforts some ninety books were translated from Arabic into Latin. Other translations were made in Sicily and elsewhere in Spain, but Toledo became the most important centre.

Among Islamic books studied or translated in Toledo, there were several which discussed mechanical devices, including astronomical instruments and several types of water clock. One author who wrote on this subject was al-Muradi, and he illustrated elaborate gear trains, some with epicyclic and segmental gears.[4] It is particularly interesting to note that he was working at almost exactly the time Su Song was building his great clock in China. Indeed, one of al-Muradi's designs was for a clock driven by a water-wheel like Su Song's, but no connection between the two seems likely. A more relevant connection is with two water clocks of rather simpler design which were operating at Toledo in the 1080s.

Mechanically minded people in Europe were becoming interested in clockwork at about this time and, over the next two centuries, learned a good deal from Islamic sources. This can be seen in a compilation of extracts from Arabic authors which was written at the court of the Christian king Alfonso X in 1276. It included a design for a weight-driven clock regulated by a hollow wheel containing mercury. One can trace ideas in this which probably came from al-Muradi, and perhaps even from Bhaskara, the Indian author mentioned in chapter 2. When the first successful weight-driven clocks were built in Europe after 1300, many ideas from other Islamic devices were incorporated in them, especially with respect to the use of gear-wheels (and even sometimes epicyclic gears). The one really original new idea was the escapement device for regulating the rate at which the wheels turned and for ensuring that the clock kept time.

Innovations of a more practical, everyday kind came into Europe as a result of the Crusades. From 1096 onwards, these expeditions to the eastern Mediterranean gave Europeans new experience of fortifications and incendiary weapons. Unfamiliar foods (and recipes learned from a translated Baghdad cookery book) stimulated the adoption of new food-processing technologies, notably for making pasta. More spices were imported, and new crops were grown, especially in Italy.

In order to understand this influx of Islamic techniques and knowledge into Europe, it has to be appreciated that European technology was already developing rapidly, with a considerable impetus of its own. People cannot adopt technologies from other cultures unless they have the skills necessary to modify, adapt and develop them to suit their own purposes. Thus the European ability to learn so rapidly from contact with the Islamic world was the outcome of previous experience of innovation in agriculture and in the use of mechanical devices. The case has been convincingly made by a number of historians (including Lynn White and Jean Gimpel)[5] for saying that the roots of Europe's later economic and technological successes are to be found in the Middle Ages, dating back at least two centuries before the period under discussion here. Viking ships, the wheeled plough and innovations in building techniques can all be quoted to demonstrate this.

However, the most outstanding feature of the European scene was the extent to which mechanical equipment and non-human energy sources were used. Water-wheels not only drove numerous corn mills, but worked mills for fulling cloth and, as we have seen, for preparing paper pulp. A new type of windmill was invented soon after 1150. These developments are not described here since they have been discussed in a complementary volume.[6] Suffice it to say that if we see the use of non-human energy as crucial to technological development, Europe in 1150 was the equal of the Islamic and Chinese civilizations. In terms of the sophistication of individual machines, however, notably for textile processing, and in terms of the broad scope of its technology, Europe was still a backward region, which stood to benefit much from its contacts with Islam.

The Mongols and gunpowder

In China, the Islamic countries and western Europe, then, the world of 1150 had three major cultures which were outstandingly 'mechanically minded'. Although Europe was the least advanced in this respect, by 1450 European technology was developing faster than that of the other countries, and in new directions. Social and political institutions in Europe had favoured 'the multiplication of points of creativity' in the many small states into which the continent was divided. Equally significant, however, was the fact that major disasters had befallen the Islamic world and China, as chapter 2 has already indicated, and had left a heritage of conservatism and caution. One notably perceptive historian of European industry compares these events to the collapse of the Roman Empire in the fifth century, a disaster which 'set European science back almost a thousand years'. Similarly, the Mongol invasions of the thirteenth century ended the classical Islamic civilization, and new 'Dark Ages' followed.[7]

Eastern Europe experienced a Mongol invasion also, but the overall effect in the West was beneficial, because the Mongol Empire provided a means of better communication with China. In 1237, Mongol armies had begun the conquest of southern Russia, then in 1241 entered Poland and Hungary in a two-pronged attack. After only a brief occupation, though, news came from Mongolia that the Great Khan Ogedei had died. He was the son and successor of Chingiz Khan, and with his death the commanders in Hungary felt uncertain of the future and drew back into Russia.

Europeans were prompted by all this to take a closer interest in happenings far to the east. Four years after the invasion of 1241, the pope sent an ambassador to the Great Khan's capital in Mongolia. Other travellers followed later, of whom the most interesting was William of Rubruck (or Ruysbroek). He returned in 1257, and in the following year there are reports of experiments with gunpowder and rockets at Cologne. Then a friend of William of Rubruck, Roger Bacon, gave the first account of gunpowder and its use in fireworks to be written in Europe.

A form of gunpowder had been known in China since before AD 900, and as mentioned earlier, in 1040 some recipes for gunpowder mixtures appeared in a printed book. One mixture was for making incendiary weapons and another gave a mild explosive. Rockets were invented in China before 1150, and a gunpowder formula which produced violent explosions was known a century later. Much of this knowledge had reached the Islamic countries by then, and the saltpetre used in making gunpowder there was sometimes referred to, significantly, as 'Chinese snow'. The Mongols did not have gunpowder weapons in Hungary in 1241, but acquired them from the Chinese soon after. Thus European travellers in Mongolia such as William of Rubruck could have learned about this well-established branch of Chinese technology there. Other Europeans had discovered that the Islamic nations possessed gunpowder weapons through the painful experience of being attacked with incendiary devices and grenades, as well as with petroleum-based 'Greek fire'. In 1249, for example, gunpowder was used against Crusaders in Palestine with particularly terrifying effect.

During the 1250s, the Mongols invaded Iran with 'whole regiments' of

Chinese engineers operating trebuchets (catapults) throwing gunpowder bombs. Their progress was rapid and devastating until, after the sack of Baghdad in 1258, they entered Syria. There they met an Islamic army similarly equipped and experienced their first defeat. In 1291, the same sorts of weapon were used during the siege of Acre, when the European Crusaders were expelled from Palestine.

The situation in China during this period was that, although the Mongols had conquered the north in 1227, the Song dynasty still held the central regions south of the Yellow River. Gaining control of this territory was to occupy the Mongols for more than twenty years, from the 1250s until 1276. The latter part of this campaign entailed several long sieges of Chinese cities in which a new type of trebuchet was used. It was of Arab design, with counter-weights which enabled it to throw missiles (including explosive bombs) much further than Chinese trebuchets could manage. A group of five men from Iran and Iraq who were serving as engineers with the Mongol army are likely to have played a part in its introduction.

The presence of these individuals in China in the 1270s, and the deployment of Chinese engineers in Iran, mean that there were several routes by which information about gunpowder weapons could pass from the Islamic world to China, or vice versa. Thus when two authors from the eastern Mediterranean region wrote books about gunpowder weapons around the year 1280, it is not surprising that they described bombs, rockets and fire-lances very similar to some types of Chinese weaponry. One of these authors, Marcus Graecus, is known only from a Latin translation (with many technical terms still in Arabic), whilst the other was a Syrian named Hasan al-Rammah.[8]

Although several types of gunpowder weapon are described in these books, none of them can properly be described as either a hand-gun or a cannon. This is important because, in both Chinese and Arabic languages, terms which were later applied to guns were regularly used in connection with earlier types of weapon. Thus many accounts of thirteenth-century warfare in China, Syria, and North Africa can be misunderstood as referring to guns. The possibility of misunderstanding is particularly great when we hear of Arab armies with siege engines which could throw 'small balls of iron'. If al-Rammah's book is representative of the state of the art in 1280, then vague references to guns and cannon balls going back as early as 1204 in North Africa and the 1250s in China must refer to earlier types of gunpowder weapon.[9]

Even if no guns had been made prior to 1280, however, the two key elements in the necessary technology were already present: high-nitrate gunpowder, and cylindrical metal barrels in which it could be ignited. It will be recalled that early gunpowder mixtures, used to make incendiaries, rarely contained more than 50 per cent saltpetre. A high-nitrate powder capable of exploding with sufficient force to propel a missile must contain about 70 to 75 per cent. This was probably achieved in China at least by the time of the siege of Kaifeng in 1126.

The other main element in the development of guns originated as part of the 'fire-lance', a weapon used in China, and also, by 1260, in Islamic armies. It consisted of a tube made of bamboo, wood or metal mounted on the shaft of a lance, and was filled with a mixture of gunpowder, toxic chemicals, lead pellets and pottery fragments. When ignited, it spouted flame and sparks for perhaps five minutes, rather like a Roman candle.

This is one of the devices which several authors rather misleadingly describe as a crude gun.[10] The difference between a fire-lance and a gun is that the former could utilize a relatively low-nitrate gunpowder which was ignited at the mouth of the barrel. In a gun, by contrast, a stronger gunpowder was ignited via a touch-hole at the base of the barrel, and would explode suddenly rather than burning for several minutes, expelling the entire contents of the barrel.

It appears that Chinese soldiers under Mongol command were equipped with weapons which had made the transition from fire-lance to gun in 1288. A bronze barrel has been found at the site of a battle which took place that year in Manchuria. The barrel was designed to fit on the end of a wooden shaft which probably resembled the shaft of a fire-lance. But in contrast to the fire-lance, it was designed for an explosion at the base of the barrel, not slow burning from the mouth, in that it has thicker walls and a touch-hole in the region where the explosion would occur (figure 14).

The thickening of the walls of gun barrels around the point of explosion became a distinctive feature of Chinese guns. In another early type, designed for mounting on a bench or frame in a fixed fortification, this thickening became even more pronounced. The result was a gun barrel which looked externally like a metal vase or bottle. It is striking that the earliest European illustration of a gun, drawn in 1326 or 1327, shows a barrel of precisely this type, mounted on a bench and firing an arrow. This bottle-shaped barrel is very reminiscent of Chinese practice, and in China too, arrows were fired from guns before spherical shot became standard (figure 15).[11]

Because of the early date of the European bottle-shaped gun, and a lack of Chinese evidence until a little later, it was once thought that guns were a European invention, made in response to the introduction of gunpowder, and that any similar Chinese weapons came later. With the discovery of gun barrels in China dating from 1288 and 1332, this view is no longer credible. So we would do well to take note of the Chinese

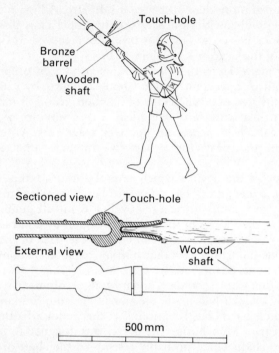

Figure 14   An early European hand-gun (top) and two older ones from China (below), all of the type derived directly from the earlier 'fire-lance'.

The barrels of the two Chinese guns survive, but the wooden handles shown here are reconstructed. The bottom gun dates from *c*.1288, and the other from 1351.

(Illustration developed from photographs and drawings given by Needham, *Science and Civilization in China*, volume v, part 7, by permission of Cambridge University Press.)

tradition that knowledge of guns passed from China to Europe via Russia.[12] With Mongol rule in southern Russia, and Mongol hegemony over Moscow and Novgorod to the north, there were a few Chinese living in these latter cities.[13] At the same time, there were trade routes from Novgorod to the West via the Baltic. The plausibility of this as the route by which knowledge of guns travelled is enhanced by the discovery of small and very early bottle-shaped barrels in Sweden.

This route bypassed the Islamic world, where no example of guns of this precise kind has been found. However, there was much experience of other gunpowder weapons in Syria, in the remaining Islamic enclave in Spain, and in the Mamluk kingdom of Egypt, and it is quite clear that

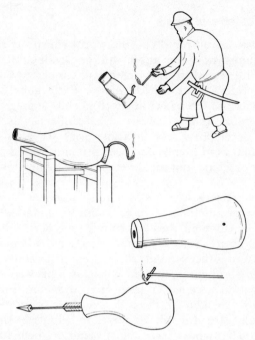

Figure 15   Early guns of the bottle-shaped type, showing similarities between Chinese and European examples.

The two top drawings are based on Chinese illustrations and show guns fired via a fuse. Below is a gun barrel found in Sweden which measures roughly 300 mm long. The bottom drawing is from the earliest known European illustration of a gun (1327), which is shown firing an arrow.

(Redrawn from Needham, *Science and Civilization in China*, volume V, part 7, by permission of Cambridge University Press.)

guns and indeed cannon were in use in Islamic areas of Spain by the 1330s.

Elsewhere in Europe, the development of cannon was very rapid. A document drawn up in Florence in 1326 shows that the city authorities were acquiring 'metal cannon' and iron shot as if they were already commonplace. That is significant, because guns large enough to be called cannons had yet to appear in China. So it may be true that the cannon, if not the gun, was initially a western innovation, and that it had developed from the smaller guns which seem to have been brought to Europe from the East.

The diffusion of technology

New technology spread rapidly in the world which the Mongols dominated, yet almost no innovation can be ascribed to the Mongols
themselves. With regard to warfare, they continued to think in terms of
horses whose riders were skilled archers. Early guns were much too
clumsy and inaccurate to use with cavalry. Thus among the Mongols in
Russia and China, and in cavalry-oriented Persia also, there was a
reluctance to develop and use firearms. Siege weapons throwing gunpowder incendiaries or bombs had been useful to Mongol armies during
their conquest of China, but the development of guns slowed under their
rule.

One other characteristic of the Mongols' lifestyle was that few of them
had experience of administration or engineering, and so to govern the
territories they conquered, they had to rely on others.

In Iran, they were generally content to leave administration in the
hands of Persian officials, but there were occasions when Chinese
technicians were brought in, notably when attempts were made to restore
neglected irrigation systems. An innovative arch dam some 26 metres in
height was built near Qum in the course of these restorations. In Russia,
Chinese officials were probably employed to help with census-taking and
tax-collecting. This would have taken some of them to Moscow and
Novgorod, which had to pay tribute to the Mongol khanate. In China
itself, Qubilai Khan was very wary of reliance on Chinese administrators,
and recruited foreigners for senior posts, including Persians, Arabs,
people of Turkish background, and the Italian Marco Polo. Attempts
were even made to persuade the pope to send a hundred trained men
from Europe.[14]

To the extent that the travels and work of such people led to the
communication of knowledge and technique from the West into China,
or vice versa, we may speak of a *transfer of technology*. In most cases,
however, we only have sketchy details of the people concerned. Thus the
appearance of new techniques in Europe which had previously been
known only in China or Islam can only be vaguely associated with trade
or supposed movements of craftworkers. Paper-making is one of the few
examples where the evidence is clear: skilled Chinese introduced the
technique at Samarqand and Baghdad, whilst later, Islamic workers
brought it into Spain (see table 2). Another clear case is the transfer of
specialist skills in glass-making from Islamic Syria to Venice following a
treaty between a local ruler and the Doge of Venice in 1277. One of the
raw materials used to make the glass was known in Syria as 'al-Qali'.
This was potash, and the fact that the word passed into European
languages as 'alkali' is symptomatic of the transfer of a considerable

body of chemical knowledge from the Islamic world,[15] through books as well as in connection with processes such as glass-making.

In examples such as this, the idea of *transfer of technology* works well, signifying that techniques and knowledge were moved wholesale from one regional and cultural setting to another. However, the deficiency of this phrase is that it implies a process in which the recipients of a new technique passively adopt it without modification. The reality is that transfers of technology nearly always involve modifications to suit new conditions, and often stimulate fresh innovations. The obvious example is that the transfer of gunpowder recipes and some primitive hand-guns from China stimulated the invention of the cannon in Europe, probably in the decade 1310–20. Thus the invention of the cannon can be seen as the outcome of a dialectic or dialogue between the eastern and western parts of the Old World, triggered by the transfer of gunpowder and early firearms technology from China.

In such instances, transfer of technology is just part of a more complex process, which also includes inventive responses from recipients of the technology. Sometimes, the transfer itself is of a minimal kind. This is the case when quite vague information from another country, or an unusual artefact, is sufficient by itself to stimulate innovation in the recipient country. Some commentators describe this as 'stimulus diffusion'. However, the concept of a technological dialogue or dialectic, or an 'inventive exchange', is comprehensive enough to cover all these ways in which ideas may spread and develop.

In discussing the dissemination of technology, we should also be aware of the possibility of *independent invention*. The windmill which first appeared in Europe soon after 1150 was so different from the Persian windmill shown in figure 5 that it must be regarded as an independent invention. The same is likely to be true of other inventions which Joseph Needham[16] claims were transferred from China, including the blast furnace and many aspects of printing. Broad similarity between techniques is not sufficient, in the absence of other evidence, to establish a connection. Only if similarities extend to idiosyncracies or highly specialized features, such as the peculiar bottle shape of early guns in both Europe and China, may we reasonably claim that technology has been transferred.

Another example belonging to the decades around 1300 is the silk-reeling machine of the kind shown in figure 8. Even as late as the nineteenth century, silk was reeled in Europe on machines with the same sort of reel, a similar frame, and a heated bath of water for the cocoons, as we shall see in chapter 6. Moreover, the last part of the route by which this technology was transferred into Europe is fairly clear. In 1050, there was a silk industry in areas ruled from Constantinople, and also in parts of the Islamic world, including Tunisia, Sicily and Granada (in Spain). But there was no silk industry at all in the western, Christian part of Europe. Just before 1300, we hear of silk manufacture at Lucca, in northern Italy, and it appears that knowledge of relevant techniques had been obtained from Sicily. Soon there were elaborate silk 'throwing' machines at workshops in Lucca and Bologna. But again, there is evidence that a process of adaptation and 'dialogue' had taken place. Whilst reeling machines are likely to have been similar to Chinese types wherever they were used in the Mediterranean region, the 'throwing machine' used at Lucca for twisting the thread was an unusual device with a circular frame for which there was no precedent in China. So although some silk technology had been transferred from China with minimal modification, some had also been invented in the West.

One other question to raise about the reception of Asian techniques in Europe is not just the fact that they stimulated fresh innovation, resulting in the cannon and (if it is not an Islamic invention) the silk-throwing machine. Beyond that, one must enquire about the extraordinary vigour with which some inventions were developed. A partial explanation relates to social institutions. Europe was a collection of small states and self-governing cities. Some Italian towns not only recruited their own volunteer soldiers but commissioned improved types of weapon. What amounted to an 'arms race' in Italy after 1300 led to improvements in crossbows, plate armour and guns.

However, while an arms race and commercial competition may go a long way to explain the energetic inventive activity of the period, there was also a quality of imagination in what was achieved which suggests stimulus from other sources also, including perhaps cosmological and religious ideas. Some projects evidently had a strong appeal for the people working on them, possibly because they were tied up with ideas of that kind. In a previous generation, this had been very much the case for the people who built the great cathedrals of Europe, with the complex technical problems of construction which that involved. One can well understand how the cathedrals embodied ideas about the relationship of earth and heaven, and so became symbols expressive of the dreams of their builders.

Similarly, the very obvious symbolism of a clock as a direct represen-tation of the sun and stars is related to the great appeal which these mechanisms had for some people. It helps to explain the immense effort devoted to the development of weight-driven clocks in Europe from about 1300.

The question which next arises is whether any of the same symbols and imaginative themes extended to the other great technological project of the period, the development of guns. Claims have been made about the sexual symbolism associated with the physical appearance and function of guns, and it would be possible to talk about sexual symbolism associated with machines and cathedral design also. But our modern awareness of this aspect should not lead us to overlook the strong imaginative appeal of cosmological symbolism. To understand this it is necessary to think about the noises made by guns rather than their phallic shapes, and it is relevant also to note responses to guns in China. Some weapons there were given names which explicitly referred to cosmic or elemental forces. There was an 'overawing wind-fire cannon' and several types of 'thunder-fire cannon'. There was also a 'magically efficient' fire-lance, and ideas about magic, denoting the unnatural and the powerful, are associated with the use of gunpowder in Europe also.

Barrels for cannon were cast from almost the same bronze alloy as church bells (though often containing less tin), using the same techniques. Murray Schafer, an expert on acoustics, suggests a parallel symbolism. Church bells might be rung to warn of the start of a battle and to celebrate its end; cannon might also be fired to mark the same events. More than that, bells and cannon were both means of making noises which Schafer describes as 'divine thunder', meaning that they had cosmic overtones. On the other hand, in ringing out the hours which regulated the life of a town, bells lifted 'all things unto a sphere of order', whereas cannon symbolized disorder.

None of this would matter very much except that guns were often used in a way in which the symbolism associated with the noise they made was more important than any material damage they did. Sounds made by a cannon being fired could intimidate an enemy and give a sense of power and confidence to an attacker, as drums had also done, even when guns were so inefficient and difficult to aim that their practical use was limited. At first, the main practical use of cannon was as siege weapons, for they were good at demolishing castle walls. But when battles took place in the open field, it is arguable that 'artillery had only a psychological effect'. In China, guns were used chiefly at the beginning of a battle to impress the enemy, and then the real fighting was done with the more accurate and deadly crossbow. Except in sieges, then, the role of the new weapon was to make a lot of noise, because noise is associated with the symbolism of power. Thus, 'if cannon had been silent, they would never have been used in warfare.'[17]

The fourteenth-century trauma

Firearms would have developed less rapidly had it not been for frequent warfare which stimulated demand and extended experience of their performance. This was true in China as well as in Europe, being associated with rebellion against Mongol rule which coalesced into a coherent campaign around 1356, and which succeeded in expelling the Mongol emperor in 1368. It is no coincidence that the earliest large cannon found in China date from the former year, and were made for a rebel faction, because whilst the Mongols had neglected this branch of technology, various unofficial groups had continued to innovate. The first European cannon may have been made about forty years earlier, and there was further development during the Hundred Years' War from 1337.

Along with the devastation of war, bubonic plague spread across Asia and Europe during this period. The first outbreaks may have occurred among Mongol soldiers in the mountainous southwest of China, from where the epidemic spread widely through the country in 1331. A Mongol army was again involved when the disease appeared in southern Russia in 1347, and from there progressed through Europe as the notorious Black Death. Then China experienced a further outbreak in 1353. The combined effect of plague and war was catastrophic. A census in China recorded a population of 123 million in 1200, but there were only 65 million people in the same territory in 1393. In Europe, it is thought that population declined in the middle fourteenth century by 20 to 30 per cent.

In both regions, the dreadful loss of population left psychological scars as well as economic disruption. A common reaction was to seek reassurance in traditional values. Religious mysticism and deep penitence for imagined wrongdoing are reported in Europe, whilst in China, an introspective neo-Confucianism surfaced later. Often there was suspicion of alien ideas and minority groups. Jews were persecuted in Europe, and in China, foreign Moslems were already suspect in their role as officials appointed by the Mongol government.

In Europe, these depressing trends did not have a prolonged effect on intellectual life and technical innovation, particularly in Italy. Developments in China took a different course, however. Attitudes varied between people who acquiesced to Mongol rule, and those who withdrew from public life, appalled by the way non-Chinese officials and local collaborators carved out large estates for themselves. Dispossessed peasant farmers were sometimes made to work on these estates as forced labour, or sometimes were left to starve. Former officials who felt disgust at this tended to re-emphasize the traditional Confucian concern for the welfare of the agricultural community on which Chinese life was felt to depend. Such men might often be conservative in their view of trade and technology whilst retaining an interest in farming and basic production.

Expulsion of the Mongols and establishment of the Ming dynasty in 1368 did not immediately mean that austere Confucian attitudes prevailed, however. The success of the anti-Mongol campaign had depended on guns and ships, and after the Ming capital was settled at Beijing, coastal shipping was required to transport grain supplies from the south. For this and other reasons, government shipyards built over 2,100 sea-going vessels in the sixteen years from 1403.

The early Ming period can thus be seen as a time of active technological innovation, beginning with the cannon made in 1356. Although guns in their simplest form had been invented in China, and then Europeans took the next step and developed large cannons, the Chinese cannons of 1356 had the first barrels to be made of cast iron rather than bronze. Other evidence of renewed vigour in the iron industry after its decline during the Mongol period was to be seen in the reconstruction of several iron-chain suspension bridges in the southwestern provinces. There were other innovations in the shipbuilding programme of the years after 1403, and in engineering works on the Grand Canal in 1411, which improved the water supply to its summit level.

However, this phase of technological innovation encouraged by government proved to be short-lived. Once grain transport to Beijing was by canal, it could be argued that a large navy was no longer necessary. Navies were associated with disastrous adventures and destabilizing, foreign trade. Moreover, some of the officials responsible for the Chinese navy were Moslems, notably the famous admiral Zheng He (Cheng Ho), whose father had even been on a pilgrimage to Mecca. Such people were all too reminiscent of the foreign administrators appointed by the Mongols. After 1415, shipbuilding resources were diverted into the construction of canal boats, and then in 1419 official shipbuilding stopped completely. Finally, in 1435, overseas voyages and all but a minimum of naval activity ceased, so there was no further incentive for technical innovation in shipbuilding.[18]

Other technologies which had traditionally been sponsored by the state were restricted either by tight controls or by lack of interest, and there was little further development in gun-making, or in mathematics and clockwork.

These attitudes are difficult to explain in detail, although the traumas of plague and Mongol occupation in the previous century make some policies understandable. It is significant also that the Mongol threat had not disappeared. Large numbers of troops were maintained along China's northern border, but even so in 1449 the emperor was captured by Mongol and Tartar armies when visiting the area. It is not surprising, then, that much effort was devoted to strengthening and rebuilding the Great Wall along the line of much older and probably more limited fortifications. This enormous project may have been another negative influence on technological development, since it absorbed massive resources without stimulating much innovation.

This more conservative approach to technology can be associated with fairly deliberate decisions which were taken in China from 1415 onwards, and were to be of great significance in an age when guns and ships, not static fortifications, were to be the means by which certain nations achieved a degree of world dominance out of all proportion to their populations and resources. One of the most important developments of the period was the European concept of the three-masted ship, which became the pattern for western ocean-going vessels, and (related to it) the small but versatile Portuguese caravel. In most respects, Chinese ships built before 1419 were more sophisticated, but these were being withdrawn from service before caravels were first used in voyages along the Atlantic coast of Africa.

However, it is a great mistake to assume that all aspects of Chinese technology stagnated from this time onwards, just because there were few developments in government-supported technologies such as shipbuilding. In later chapters, we shall notice many innovations in agriculture and the textiles industries, and a considerable expansion of iron-smelting.

In the thriving printing trade, for example, one major innovation was movable *metal* type, as distinct from earlier methods of printing from wooden blocks. The first place where metal type was used for printing was in Korea, perhaps as early as 1234. However, that was an isolated and tentative beginning. Only in 1402 and later is there evidence from Korea of a consistent development of metal type. New fonts, each consisting of over 100,000 pieces of type, were introduced in 1402, and then in 1420, 1434, 1436 and so on.

In China, a similar system of printing using metal came into use around 1490, after earlier experience with movable type made from other materials (including ceramics). We may suspect a transfer of technology from Korea to China, including knowledge of ink and paper qualities for use with metal type, but the evidence is not clear.[19] What we cannot doubt is that when metal type was first used in Europe about 1450, this was an independent invention. The different script used in Europe required a different approach to casting type, with greater emphasis on quantity production of rather smaller pieces. The Koreans were able to cast their bronze type in small sand moulds, but the European innovators had to develop a metal matrix for the purpose.

From some points of view, however, it was the Portuguese caravel, not printing, which had the more immediate implications. This small ship was probably developed quite deliberately for exploration of the African coast. One of the critical problems in this was that prevailing winds and currents off southern Morocco were from the north. If a ship was to return home after a voyage to West Africa, it needed to be able to make good progress against unfavourable winds. It was for this job that the caravel was such an advance.

The famous voyages of exploration which were sent out from Portugal by Prince Henry the Navigator were cautious and systematic, based on carefully recording and mapping navigational data. In 1443, the first contacts were made with Negro communities near Cape Verde, and thought was given to establishing a trade route inland to Mali. Thirty years later, when the coast of West Africa had been fully explored and forts had been set up at strategic points, gold was being obtained on the coast of what is now Ghana. Ominously foreshadowing future exploitation, slaves were soon being sent back to Portugal and Madeira to work as agricultural labourers.

It was not until 1488 that the Cape of Good Hope was reached, and not until another decade had passed that Vasco da Gama sailed round the Cape to reach Moçambique. There he encountered Arab shipping and was able to secure the services of an Arab pilot to guide him along the East African coast to Mombasa and Malindi. From the latter port, another pilot helped the Portuguese ships across the ocean to South India. The irony is that Chinese fleets had visited these ports sixty years before but had now been totally withdrawn, and the Islamic naval challenge to the Portuguese, when it came, was ineffective. Western Europe's new route to Asia was thus open. In the meantime, other explorers had been crossing the oceans but, with less careful research than Prince Henry's, believed that they had reached Asia by sailing west across the Atlantic. The consequences of this and other transatlantic voyages will concern us in the next chapter.

4 Conquest in the Americas, and Asian trade

Independent invention

The reality of America dawned on Europeans only slowly. Scandinavian (Viking) settlers in Greenland had visited it regularly to collect supplies of timber from AD 990 until just before 1400. Portuguese, Danish and English seafarers followed up information from Scandinavian sources in voyages to Labrador and Newfoundland in 1476 and 1497. Meanwhile, sailors from Portugal may have glimpsed Brazil, and in 1492 Columbus travelled as far west as Cuba, but thought he was in the 'Indies'.

As to the peoples of this 'new world', Inuit (Eskimos) had been encountered in the north and Columbus had met some islanders, but the possibility of highly developed civilizations in America was hardly grasped before Cortes arrived in Mexico in 1519, marvelling at the splendours of Tenochtitlan, but horrified by the human sacrifices which took place there. What perhaps is more striking to us is that the peoples of the new world did not have iron tools (except for a very few of meteorite iron) and did not use the wheel, yet had developed elaborate 'technology complexes' for basic survival, varying greatly from one type of environment to another. Of these, the most specialized was that of the Inuit in the Arctic, and there were other hunting and fishing communities living in very varied conditions. However, it was the agriculturists of the Americas who most impressed Europeans in the sixteenth century. The two features of their farming which were most characteristic were the crops they had domesticated and some of their techniques for water conservation.[1]

One particularly striking practice was the construction of artificial islands in freshwater lakes, as developed first in the Maya culture before AD 1000, and then from 1325 used by the Aztecs around Tenochtitlan (now the site of Mexico City). These islands or 'raised fields' were formed by digging canals around narrow strips of swampy lake-margin land. Soil levels on cultivated plots were slowly built up as silt from the lake was spread on them periodically to maintain fertility. Near Tenochtitlan, a regular grid of canals developed with rectangular gardens in between. A typical plot might be 100 metres long but only about 6 metres wide, so the whole area could easily be watered from the canal-side. Under these conditions, two or three crops could be grown each year.

This is an unusual adaptation to local conditions, but other American agricultural systems have close parallels elsewhere. In the dry landscapes of what is now Arizona, fields were sometimes enclosed with low banks to retain all available rainwater, and small dams were built to hold back floodwater during storms. At one site, Point of Pines, archaeologists have found evidence of these techniques as they were practised around AD 1000. Similar earthworks were used for water conservation in parts of Arabia and Africa (both south and north of the Sahara Desert).

The Maya civilization of Central America is famous for its great temples, its system of hieroglyphic writing and its sophisticated calendar, but for a long time its agriculture was thought to be based on cultivation of maize and beans, with land left fallow for long periods when soils became exhausted. However, archaeologists have now shown that when the Maya culture was at its height, between AD 600 and 900, there were more people living around the major temples than such basic farming methods could have fed. Moreover, the remains of engineering works have been found, including drains for irrigation or drainage. Agricultural technology varied from area to area, according to ecological conditions. On low-lying land that was seasonally flooded, raised fields were created like the water examples at Tenochtitlan. On well-drained land whose natural vegetation was rain forest, trees producing high yields of nuts were regularly planted, and it is likely that manioc (cassava) and vegetables were grown under the trees in a manner comparable to the 'multistorey farming' systems of Indonesia mentioned in chapter 2. The trees protected the soil from erosion and helped maintain fertility through leaf fall, so soils were not exhausted as under maize farming. The result has been compared to an 'artificial rain forest', and food production could have been ten times greater than from alternating maize plots and fallow land.[2]

In Peru, irrigation based on the rivers Chicama and Moche developed before the rise of the Incas, with long canals fed from diversion dams. One canal was 110 kilometres long, and had been so carefully surveyed that a uniform gradient was maintained throughout a descent of 1,200 metres. One historian compares this high-quality engineering with what was achieved in the Islamic world at about the same time, and comments that it 'illustrates how basic a technology irrigation is'.[3] The point here, as with water conservation in Arizona and closely comparable techniques in Africa, is that people with comparable environmental problems will often come up with similar inventions quite independently. But for Joseph Needham, independent invention is not an adequate explanation for similarities between terraced cultivation in Peru and in China. Nor does it account for similarities he notes between rope suspension bridges in the Americas and China, or parallels between the Maya calendar and its Chinese counterpart. He therefore argues for some kind of early contact between Asia and America.[4]

Perhaps the most striking example here is mulberry-bark paper. Used before linen-rag paper was introduced in China, this fabric was made into a form of cloth in parts of Indonesia until modern times, and spread from there to some of the Pacific islands. In Central America, a bark fabric of basically the same kind was used for writing the Maya hieroglyphic script, perhaps from as early as AD 700. This geographical distribution suggests a step-by-step diffusion of the techniques from one island to another, and thence to America, not direct voyages from China. Transfers of agricultural technology and boat-building methods from Indonesia eastwards certainly occurred, and it is very likely that bark paper travelled the same route. What contact was made with the Americas at the end of the line remains problematic, however.

Biological resources

A more instructive comparison between Central American and Chinese civilizations has been suggested in an important book by Ester Boserup, *Population and Technology*. She points out that around AD 900, population densities in the two regions were unusually high by comparison with other parts of the world at that date. This makes the two civilizations highly pertinent to Boserup's view that, historically, technological development was always strongly influenced, even determined, by population density. Thus the Maya culture at its height and China at the inauguration of the Song dynasty were both practising very productive forms of agriculture, with various engineering techniques for water control well developed. The crucial difference was that the Maya culture lacked iron tools, the wheel and animals capable of pulling ploughs. Ploughs were not very relevant to forms of agriculture practised by the Maya, but Boserup's point is that no options were open to them for offsetting heavy labour requirements by using machines, animal power or improved implements. The relative absence of domestic animals may also have meant that extra labour was needed for manuring the land. In terms of Boserup's concept of 'technological levels', defined in terms of tool use and energy resources, this was a very low-level technology. A biological means of offsetting heavy labour demands was already in use, however. Some of the plants which had been domesticated in the Americas were more productive of food per hectare than any Asian crop (except some kinds of rice), especially when grown in raised fields or under artificial rain forest conditions. In tropical areas, manioc, maize and also ground-nuts were important, whilst potatoes were the key food crop in the highlands of Peru. It was such crops which made it possible for population densities to rise without imposing unrealistic work burdens on farmers, despite the lack of labour-saving equipment.

It would be wrong to think that all the problems of supporting a complex civilization had been solved, however. In Maya territory, there was a sharp decline in population some time after AD 900 or 950. Much land reverted to forest or brush, and a smaller number of people supported themselves by a reversion to simple maize cultivation and fallow.

In Peru, the Inca Empire which developed five centuries later was of a very different character. Written records hardly existed, yet adminis-tration was highly efficient. There were remarkable stone buildings, and the llama played a larger part in the economy than any other American domestic animal, producing meat and wool as well as being used to carry loads on its back. Copper, gold, tin and silver had been mined in the Andes even before the Inca Empire arose, and were sometimes used in surprising ways. For exampe, 'tumbaga' (a gold/copper alloy) was made, which was almost as hard as bronze.

The arrival of Europeans on the mainland of Central America in 1513, and more permanently in 1519, led to disaster unparalleled in the history of any major civilization. The combination of military conquest and epidemic disease had appalling results, not least because the people had no prior experience of smallpox (which reached Mexico in 1520) or measles (in 1530–1), and hence no natural immunity to either disease. The first smallpox outbreak occurred in an Aztec force which had just compelled a Spanish retreat from Tenochtitlan. It paralysed all further resistance. Death rates were very high, and people were demoralized by superstitious explanations of the disease. The loss by many people of their 'will to live' led to suicide and the neglect of newborn babies, so that mortality was higher than need be. Mexico is thought to have had a population of 25–30 million in 1500. By 1568, only about 3 million remained and numbers were still declining.[5]

Such a catastrophic decline in population understandably led people to lose confidence in their own culture and institutions, with the result that Spanish language and religion quickly became dominant. There was also a loss of indigenous skills. However, in regions to the north of Mexico, which Europeans had yet to penetrate in force, there were a few peoples whose culture continued to develop independently, but now with some transfers of technology from Europe. The most important introduction was the horse, then later iron tools and firearms.

With declining skills in Central America, and very limited tools even before the conquest, it may appear that there was almost no scope for a transfer of technology to the rest of the world. However, the heritage of American domesticated plants was of enormous significance for Europe, Asia and Africa, stimulating something of an 'agricultural revolution' in one country after another, and helping to support an accelerating growth in world population from the middle of the seventeenth century. As one historian has said, the 'contribution of the indigenous Americans to the future' was not made through technical skills and institutions, which all but disappeared, but arose from 'obscure, unrecorded discoveries by primitive cultivators who first discovered how to exploit the ancestors of maize [and] potatoes'. They had 'unwittingly made a huge addition to the resoures of mankind'.[6]

Apart from food crops, which included groundnuts, manioc (tapioca), chilli peppers and tomatoes, as well as two kinds of potato (botanically unrelated) and maize, the American heritage also included medicinal plants and tobacco. Substances which were exploited by European physicians included sarsaparilla, a root from Mexico and Peru from which a tonic was prepared in Europe from the 1570s, and cinchona bark from the Andes, used to treat malaria from the seventeenth century and ultimately the source of quinine. There were also industrial crops whose importance was still largely in the future. Logwood dyes were exploited from the seventeenth century, but the significance of rubber was very slight until much later.

Some of the American food crops offered such advantages that they spread remarkably fast. In 1593, for example, there were poor harvests and food shortages in the Chinese province of Fujian, and its governor sent a mission to the Philippines in search of plants that might improve food production in subsequent years. They returned in 1594 with a new root crop, the sweet potato. This was a plant which Columbus himself had brought from the Americas to Spain, and which the Spaniards had introduced into the Philippines. It grew so successfully in Fujian that it soon became the most important root crop in South China, and spread from there to Taiwan (soon after 1605) and to Japan (1698). The so-called Irish or white potato filled the same role in Europe, but was not brought from America until about 1570.

The Mongols of the seas[7]

Although it seems unlikely that Chinese ships ever visited the Americas before this period, or indeed for a long time after, there were certainly some important Chinese voyages of discovery in the fifteenth century. During the great shipbuilding effort of the years between 1403 and 1419, several exceptionally large vessels were constructed, each perhaps 100 metres long and specifically designed for extended voyages. They were used for a series of expeditions into the Indian Ocean under the direction of the admiral Zheng He. The purpose was to establish diplomatic relations with the rulers of other nations, to encourage trade, and to collect information on navigation, geography and natural history. The ships visited virtually every Asian country with any sort of coastline, using first Malacca, then a Sri Lankan port and then Calicut in South India as bases. They carried armed soldiers, but in most places good relations were established by exchanging gifts for 'tribute' despatched to Beijing. Only rarely was armed force used, most notably in Sri Lanka. However, a monument erected there in 1411 testifies to a friendlier relationship also – it has inscriptions in the Chinese, Tamil and Persian languages recording gifts to local religious institutions brought on one of these voyages.

The Chinese fleet also explored the coastline of East Africa, which had been trading indirectly with China through Arab and Indian intermediaries. China imported African ivory, and its exports of porcelain sometimes found their way to Zimbabwe through middlemen in Gujarat. The last of Zheng He's voyages across the Indian Ocean took place between 1431 and 1433. Soon after his return, the political changes in China discussed in the previous chapter led to the cancellation of all further voyages.

Had the Chinese still been patrolling the Indian Ocean when the Portuguese arrived, one can only speculate what might have happened. The decision to withdraw the Chinese fleet was a momentous one, not only for what it portended with regard to China's own development, but also for what it meant in world affairs to have the 'door left open' into the Indian Ocean. The ruler of Calicut, the first Indian port reached by Vasco da Gama in 1498, soon recognized the danger, and the Islamic world quickly saw that its trade was threatened. In 1507, a large fleet set out from ports on the Red Sea to confront the intruders. In the next year, with Indian allies, they forced the withdrawal of some Portuguese ships, but in 1509 were disastrously defeated by a small group of Portuguese vessels off the west coast of India.

Although the Red Sea force was equipped with guns, its commanders had not rethought their tactics to take account of these weapons. Thus they were still using galleys with the aim of ramming and boarding enemy vessels. The Portuguese had some galleys also, but chiefly depended on the manoeuvrability of their sailing ships to keep their distance whilst using gunfire to destroy the attacking vessels.

In seeking to understand what gave Europeans their power to dominate the Indian Ocean, it is useful to make a comparison with the Mongols, whose advantage in land battles lay in great mobility (on horses) combined with an effective weapon (the compound bow). Their bows and stirrups may seem simple technology, but few other nations were so well equipped, even though in other respects Mongol techniques were cruder than those of the countries they conquered. Europeans were in roughly the same position at sea. Their ships were highly mobile, and with suitable tactics this mobility could give them a decisive advantage. Thus Asia was subjected to something like another Mongol conquest, but at first limited to ports and shipping. Arab or Asian ships on long voyages were soon being forced to pay what amounted to protection money or risk having their cargo seized. Alfonso d'Albuquerque captured the port of Goa in 1509–10 and made it the main Portuguese base on the west coast of India, and then in the following year took the independent Malay port of Malacca. The only lasting success achieved by Islamic navies was to keep the Portuguese out of the Red Sea, where they attempted to establish a base in the 1520s, and to restrict access to much of the Persian Gulf.

To sum up, then, what the stirrup and bow were to the Mongols, technical advances in ship design were to the Europeans. The key points were the strongly built hull, the stern-post rudder, and the mix of different types of sail on the three masts which were now standard. Whilst the main mast was always square-rigged, the mizzen mast (nearer the stern) usually carried a triangular lateen sail. The foremast would carry more square-rigged sails, and there were also triangular jib sails stretched between this mast and the bowsprit. This combination of different types of sail could be deployed to make use of winds from almost any direction, so that it was possible to sail 'close to the wind'. A ship so equipped had great advantages over vessels which were more restricted in their requirements for a favourable wind.

Here it is worth mentioning again the Chinese ships which had been withdrawn from the Indian Ocean, since they were possibly even more advanced vessels. They were certainly larger (figure 16), and although there is some uncertainty about their dimensions, it seems likely that they displaced well over 1,000 tons compared with about 300 for Vasco da Gama's flagship. They had robustly built hulls, and of course the stern-post rudder had been used by the Chinese long before its introduction in Europe. Their sail plan would have been quite different, and it is difficult to comment on how close to the wind they could sail. Arab and Indian ships were usually similar in size to European vessels but with lighter construction. Thus they were vulnerable to cannon fire.

Figure 16   Fifteenth-century ships from China and Europe.
These ships could never have been seen side by side because the great vessels of Zheng He's fleet had been scrapped before European boats entered the Indian Ocean, and no more of comparable size were built. There has been controversy about the Chinese vessels' dimensions, with estimates for length at the keel mostly over 100 metres. This drawing does not represent such an enormous size. If the three-masted Spanish ship with a lateen mizzen sail is 30 metres long, the corresponding measurement for the Chinese vessel is 84 metres.
(Illustration by Hazel Cotterell; for evidence of the size of Zheng He's ships, see Needham, *Science and Civilization in China*, volume IV, part 3, pp. 480–2.)

Shipbuilding technology

One key point affecting the strength of ships' hulls was the use of iron nails in both European and Chinese ships, and their absence on Arab and Indian vessels. South Asian naval architecture was based on making edge-to-edge connections between planks with wooden dowels or rattan sewing. The hull was conceived and built as a shell, with ribs added later to stiffen it. By contrast, European ships were built around a framework of ribs, and Chinese hulls were subdivided by heavy bulkheads. The use of iron nails made possible a strong connection between external planking and frames or bulkheads. This produced a less flexible hull, which was a disadvantage only if ships had to be beached regularly. It was, however, a real asset in withstanding heavy seas, and proved to be less fragile when hit by cannon balls.

Indian shipbuilders were able to modify their methods of construction very quickly in the light of their experience of European vessels (perhaps because Chinese ships visiting the same ports seventy years before had already introduced new ideas). Nails were soon being used by shipbuilders in the region where Vasco da Gama had first arrived by 1501, and when Albuquerque captured Goa for the Portuguese in 1509, he found large quantities of nails in store. About the same time, a 'galleon' built for an Arab merchant which incorporated many European features was launched from a Gujarat shipyard.

Prior to this, iron production in India had been on a very small scale and there was apparently some difficulty in supplying nails in sufficient quantity when shipbuilders began to use them. By the 1590s, however, not only had iron output increased to meet this demand, but it was supplying the makers of guns also, and iron anchors were being manufactured in place of the stone ones previously used.

Despite their obvious merits, European ships had some serious deficiencies, and were in need of constant repair. Oak planks were attacked by 'worm', but the teak planks favoured by Indian shipbuilders lasted much better. There were also some severe problems associated with the use of iron nails, problems which became more acute as ships sailed into warmer waters. The iron nails rusted very badly in this environment, causing decay of the surrounding timber. European ships could not be used for more than two or three years in the seas around Southeast Asia, and one of Magellan's ships lasted only eight months. The Chinese were well aware of the problem and rarely sent their big ships into the area. Most of their trade was carried by boats built in the Philippines or Indonesia which had no iron fastenings.

Europeans eventually found ways of avoiding exposed ironwork, but also employed locally built boats to a large extent. The Magellan expedition used a ship built on the island of Banda in 1511. The Portuguese regularly used local vessels, and when the Spanish began colonization of the Visaya Islands in the Philippines from 1570, they came to depend heavily on local ships and Filipino shipbuilders. Indeed, as Europeans consolidated their control of Asian maritime trade, the shipyards of India and Southeast Asia became important centres for innovation and transfer of technology.

The Spanish had particularly long lines of communication since they usually sailed to the Philippines from Mexico, and thus were more dependent on Asian shipbuilding technology than other Europeans. Many of their ships needed repairs or a refit after the long voyage across the Pacific, and Spanish colonialism in the Philippines was only possible because the colonists found a shipbuilding industry already in existence which was able to service their vessels, and which they eventually took over. The people of the Visayas were then made to fell and haul timber and carve it to unfamiliar shapes as well as to make ropes and sails. Spanish galleons were being built in the area by 1587, when members of Francis Drake's expedition saw one, and about eighty years later Alcisco Alcina, a Jesuit priest, mentioned a Filipino shipwright named Polakay who had worked on about twenty large galleons.

The most obvious transfers of technology which resulted from this were European techniques for framing boats and fixing planks, though wooden trenails were used to avoid the bad effects of iron nails. Other western features adopted in local shipping were decks, new types of mast (with the local tripod mast disappearing) and the stern-post rudder (instead of steering oars). However, there was some transfer of technology from Filipino shipwrights to Spanish designers also, related to the use of local timber and methods for making dowelled connections between planks. More than that, the Spanish kept large numbers of boats of local design in their fleet, using them for military operations within the region as well as for trade. The high speed at which they could be rowed made them particularly useful for carrying despatches.

In the seventeenth century, when Spanish control of the Visaya Islands was contested by a local ruler, all the naval battles were fought with small galleys or *karakoa* of local type. However, the Spanish tended to want a heavier and larger version than was usual. The result, according to one observer, was that 'the karakoa of our Filipino enemy . . . make a mockery of ours because they are faster.'[8]

In India, many shipyards remained independent of European control, so most innovations (or adaptations of European techniques) depended on initiatives by Indian shipwrights. There was an important group of shipyards in and around Masulapatam on the eastern or Coromandel coast, where ships of 600 tons displacement were being built by 1600. Elephants were used to haul them down the slipways for launching. The other major shipbuilding area was around Surat, in the northwest, where even larger ships were built. Most of the bigger ships were designed on European lines, with three masts, but with many Indian features. Much depended on whether the customer was an Indian merchant or a European company. At the same time, boats of traditional design, even some with sewn construction, continued to be built for local trade.

Some Indian techniques were distinctly better than those of their

European counterparts. By the early eighteenth century, officials of the British East India Company were reporting that 'vessels built here of teak timber according to the manner of Surat rabet work are far more durable' in local conditions than ships sent from England. Surat rabet work was a method for making the seams between the planks of a hull with a rabet (or rebate) along the edge of one plank shaped to take the edge of its neighbour. According to one contemporary description, Surat and Bombay carpenters have 'planks let into each other, with cotton and tar' (or pitch) to seal the joint. This was not only very durable, but also eliminated the cost of caulking with oakum, the usual European method of making seams watertight, which absorbed many man-hours of labour.[9]

Partly because of this saving, the East India Company found that ships cost between 30 and 50 per cent less if built in India. Other novel features of ships constructed in India were the treatment of timber with lime to make it more resistant to 'worms' and the built-in cisterns for carrying drinking water, which took up less space than the barrels carried by European ships.

In view of later prejudice, it is striking how eagerly Indians and Europeans learned from each other. For the Portuguese this was a tradition which went back to Prince Henry the Navigator and his enthusiasm for collecting information from whoever would impart it – Danish seamen, Arab merchants, Jewish cartographers. In the Indian Ocean also there was a deliberate policy of gleaning all the information they could find about navigation and shipbuilding. Vasco da Gama learned new navigational techniques from Ahmad Ibn Majid, who acted as his pilot from the East African coast to South India, and acquired several examples of the instruments used by Arab navigators. Similarly, Alfonso d'Albuquerque, leader of several Portuguese expeditions after 1507, collected Javanese maps and sea-charts.

The dependence of Europeans on Indian and Filipino shipbuilders is thus part of a pattern in which westerners exploited Asian knowledge and skill. Once again, the analogy with the Mongols is worth thinking about, in the way they sometimes depended on Persian administrators in China and Chinese engineers in Iran.

One other problem facing Europeans in Asia was that their trade was chronically out of balance, because there were very few goods manufactured in Europe which Asians wanted to buy. European products were of inferior quality, or irrelevant to Asian needs. Guns were certainly in demand, but muskets and cannon manufactured in the Islamic countries or Thailand were often of good quality. Thus almost everything which Europeans bought in India or China had to be paid for with gold or silver, often in the form of coin. Some cash was earned by carrying goods between Asian ports in European ships, or was acquired by demanding fees (protection money) from local shipping. However, there was still an enormous deficit, because Europe could not yet compete industrially. The answer was to develop sources of gold and silver in the Americas and use it to consolidate the maritime trade.

If Asia was thus characterized by superior manufacturing technologies, Europeans did at least understand mining and were beginning to make rapid progress in the use of machines for draining mines, for hauling materials from them, and for crushing ore. Within Europe itself, the most important mines were in the mountainous regions of Germany and Eastern Europe. Here, the ores of copper, silver and lead were often found together. There was growing demand for copper in the manufacture of bronze cannon, and during the fifteenth century much effort was made to improve methods for separating the small amounts of silver from the much larger quantities of other metals. The result was the so-called 'liquation' process whereby copper ores containing silver were smelted with lead. The silver tended to dissolve in the liquid lead more readily than in copper, and could be more easily extracted from the lead. The success of this process contributed greatly to the expansion of mining in Germany, as did contemporary developments in machines.

Large shipments of gold and silver began to cross the Atlantic during the 1530s, and by 1600 Spain was probably importing about ten times as much silver as the European mines were producing. The Spanish also carried much silver from Mexico across the Pacific to their base in the Philippines. It was used to finance trade with China on such a scale that the Mexican dollar was soon a recognized currency in China's coastal provinces.

These developments were partly the result of the transfer of European technology to Mexico and Peru and its use in the expansion of silver production there. One particularly important technique was the extraction of silver from low-grade ore by amalgamation with mercury. This process had been worked out in Europe some time after the lead liquation technique, and was particularly relevant to the type of silver deposit found in Mexico. After being introduced there in 1554, it was developed further by Bartolome de Medina, and was in large-scale use by 1566. There were deposits of cinnabar (mercury ore) in Spain, so the ships which sailed regularly for Mexico delivered the necessary mercury.

It is said that the Mexican mines employed over 10,000 labourers by 1600, and since labour was also required for the estates or haciendas belonging to Spanish settlers, and the population was declining, labour shortages were inevitable. The new rulers of Mexico were reluctant to enslave the people they had conquered, but introduced a system of labour obligation to local landowners which effectively made them serfs.

In Peru, the Spanish conquest of the Inca empire was accomplished in the years 1532–5. Subsequent exploration eventually revealed deposits of silver at Potosi which were so rich that people spoke of a 'silver mountain'. It was much the largest source of the metal in the Americas during the next two centuries. Potosi was located in the Andes at an altitude of over 4,000 metres. To begin with, the native silver could be mined without elaborate technology. Soon, however, lower-grade ores were being worked in Peru, and the mercury amalgam process was introduced in 1572. At about the same time, water-powered mills for crushing the ore were being built.[10] A dam feeding a canal, and then more small dams, were built in and around Potosi to supply the water-wheels. Expansion continued until 1626 when the collapse of one dam did so much damage that the system never fully recovered.

African mining[11]

In developing the sea route to Asia, and with their American interests limited only to Brazil, the Portuguese had little opportunity for involvement in American mining. However, for several centuries much of the gold available in both Europe and the Islamic world had come from Africa, via caravans of camels which crossed the Sahara Desert (see figure 13). In due course, the Portuguese found ways of diverting part of this trade onto their own ships. On the West Africa coast, they established a fort at Elmina, west of the Volta River, because gold could be obtained by trading there. The mines were some way inland, but the Portuguese could not discover where. African miners had long taken pains to ensure that Arab traders could never visit the mines, and they kept the Portuguese at a distance also. Around 1500, the latter were told that the mines were 'deeply driven' into the ground, though it is also clear that some gold came from panning river gravels. Descriptions of more recent mines in the region (especially near Bure) suggest that many narrow shafts were sunk through the hard laterite layer near the surface to the alluvial gravels below. They were 10–15 metres deep, and at the bottom tunnels fanned out, many of them connecting with other shafts. These tunnels were only a metre high, and the miners worked in a crouched position using short-handled iron picks. Mining was only done in the dry season, because during the rains the tunnels were flooded.

Because iron tools were of critical importance for the miners, the blacksmiths who made the tools seem often to have had a controlling interest and were skilled in all aspects of mining. Some labourers in the mines may have been slaves, supplied by the Portuguese from other parts of the African coastline. At surface level, the washing of gravel from the mines to separate out the gold was done mainly by women. Merchants in the area took the gold dust so obtained without any further processing, using it as currency in many transactions. For this purpose, they had small balances and weighed out the gold dust using Islamic units of measurement. Gold exported via the traditional trans-Sahara route would sometimes be cast into ingots at Timbuctu, but often this would only be done when it got to Morocco or Tunis.

Differing estimates of the quantities of gold produced in West Africa have been put forward,[12] but it seems likely that around 1500, this region was the source of about half of Europe's supply with perhaps 500–600 kilograms being exported via the Portuguese each year, and maybe four times as much crossing the desert by camel. After 1520, of course, considerably larger quantities were coming from the Americas.

The Portuguese also found gold being traded on the East African coast. The mines were located in what is now Zimbabwe, and were cut in hard rock, not alluvial deposits. Some were no more than deep trenches which simply followed gold-bearing veins downwards. In other mines, there was a vertical shaft, up to 25 metres deep, opening out into a wider space at the bottom. There were rough steps spiralling up the sides of the shaft, and when excavation was in progress, people would stand at intervals all the way up passing baskets of broken rock from one to another. The rock was cut away in the mine by using the heat of fire to crack it, then driving iron wedges into the cracks.

Mining on the Indian subcontinent and in Indonesia was often like this, and there is some evidence that Indonesians were involved in the early days of the Zimbabwe mines, perhaps around AD 600, soon after they had colonized Madagascar. Any resulting transfers of technology were of a limited kind, however. Articles made locally from copper, iron and gold seem to be of indigenous African design. In any case, it was only from about 1200 that trade with Swaheli and Arab merchants on the coast expanded. At this time, too, there were other developments, including the introduction of cotton as a crop, with hand spinnng, and weaving using looms of the kind described in chapter 3.

The sites of over a thousand mines were found by the archaeologist Roger Summers, and no doubt there were more. Not all were in use at once, but there is evidence of the prosperity they brought in the famous ruins of stone buildings at Great Zimbabwe, and in fragments of Chinese porcelain found there. It appears that the ruler of the area acquired gold from the miners by exchanging it for cattle, then exported it in order to import porcelain and textiles for use at his court.

When the Portuguese arrived, their aim was to cut out Swaheli middlemen and Arab shippers. They established themselves at Sofala, the nearest port to Zimbabwe, in 1505, and later at Tete, a river port on the Zambezi. At first, they may have been able to obtain something like

50 kilograms of gold in a year, but the quantity soon declined. Possibly this was because of the clumsiness of their efforts to control trade. Another factor was probably exhaustion of the more accessible gold. Improved technology might have allowed more difficult deposits to be worked, but not until 1635 do we hear of a commission of European mining experts visiting the area. The export of gold from Zimbabwe continued to decline, however. Then in 1693, the Changamire rulers in the south of the country gained sufficient power to drive the Portuguese off the plateau lands where the mines were located, and gold exports ceased entirely. The only useful transfer of technology which can be associated with Portuguese activities had nothing to do with mining, but consisted of the introduction of two American crops, groundnuts and maize. African farmers in the Zambezi valley were quick to appreciate them, and maize in particular was soon widely grown.

Any list of the 'high technologies' of sixteenth-century Europe would undoubtedly include shipbuilding, mining and metallurgy. To study the state of the art in these technologies, we could arguably confine attention to Europe, and examine mine remains, business records and books such as Agricola's great text on mining, published in 1556. But that is work for another volume. Here, instead of thinking about the most elaborate metallurgical and shipbuilding expertise, it has been the aim to consider the geographical extent of technological dialogue – that is, the inventive exchange of ideas between Europe and other cultures, sometimes, indeed, involving European dependence on Asian, African or American skills.

5 Gunpowder empires, 1450–1650

Turkish ascendancy

Bronze barrels for cannon were being produced in such quantity and in so many countries by about 1500 that one historian has called this a 'second bronze age'.[1] Not only was there large-scale production in Europe, the Ottoman Empire, India and China, but before 1650 guns had been manufactured in Korea, Japan, Siam and Iran, and occasionally in many other places, notably Benin, the famous West African bronze-casting centre. Since bronze was an alloy of copper with small amounts of tin, the mining of these metals expanded greatly. Copper was exported by Japan in the East and Sweden in the West, and tin mining developed fast in Malaya.

All this can be traced back to the small hand-guns made in China before 1288 (chapter 3), which stimulated the development of large cannon, first in Europe (before 1320), then in the Islamic world (1330s) and in China itself (1356). Beyond this, the acquisition of guns by Korea during the 1370s is of interest as a well-documented transfer of technology from China. After being troubled by attacks from Japanese ships, the Koreans asked the new Ming government in China for help in acquiring weapons. The Chinese were at first unwilling to share their expertise, but when Japanese ships began to attack them as well, they at first supplied a small quantity of saltpetre and other material, and then allowed a Chinese technician to go to Korea.

The first Korean guns were mounted on ships and were used to fire flaming arrows (not shot) into attacking vessels. Whether in this form, or using more conventional ammunition, the weapons were apparently a success, and the arsenals where they were made were extended in 1404. The Siamese kingdom in Thailand probably used Chinese casting methods to make guns. One fortress captured by Europeans in the Malay peninsula was found to be armed with considerable numbers of Thai cannon.

No other countries are known to have obtained gun-making technology from China, where methods and designs were both conservative and protected by secrecy, but many nations obtained guns, or knowledge of how to make them, through contact with the Portuguese.

Another source of supply, particularly for Islamic countries, was the Turkish or Ottoman Empire. In Southeast Asia there were Moslem rulers in Sumatra who acquired Turkish guns, and Ottoman exports of firearms also went to Central Asia, India and parts of Africa. Not only was this the world's biggest export trade in guns, but some were of very high quality. In order to understand the development of this expertise, we need to remember the long-standing interest in gunpowder weapons in the Islamic world. From 1280 onwards several books discussed gunpowder weapons, and it is clear that they were not simply repeating information obtained from China or Europe, but reflected practical experience. For example, although rockets were invented in China Islamic technicians devised new types, including one that could carry a half-kilogram warhead and another which skimmed the surface of the sea to attack ships. By the 1360s, Egyptian and Turkish armies had cannon, and there were experiments with range-finding devices. With rare exceptions, historians have neglected evidence of these researches, so it has not been sufficiently clear that the Islamic world had its own sources of expertise and contributed significantly to the development of the musket and perhaps other weapons.[2]

In 1364, when the Turks began to make cannon, Ottoman rule was expanding on the European side of the Bosphorus, and we hear of Turkish armies using guns in their campaigns in 1387 and 1389. The most famous victory of Ottoman fire-power, however, was the capture of Constantinople in 1453. There are many stories about the Hungarian gun-founder who assisted the Turks, but the latter were already equipped with effective cannon before he appeared. It remains true, however, that the Ottomans recruited European gun-founders whenever they could. Moreover, as their dominions in eastern Europe increased, there were new opportunities to tap European expertise, notably at Dubrovnik, which was under Turkish control from 1526, and at mining centres in Transylvania and Hungary, incorporated within the empire during the 1540s. The mining centres improved Turkish access to the metals needed for making cannon as well as providing some silver and gold.

The fall of Constantinople convincingly demonstrated that the cannon had become a very effective siege weapon. At this time also, lighter, more mobile field-guns were being developed in France, whilst improved hand-guns were being made both in the Ottoman Empire and further west. The problem with all early hand-guns was the need to apply a burning taper or match to the touch-hole for firing the gun. It was very difficult to do this whilst simultaneously aiming the weapon. By 1450, there were various devices for holding a slow-burning fuse or match, and using a trigger to apply it to the powder. The matchlock musket developed from such devices about 1470, appearing almost simultaneously among the

Turks and in western Europe. Although Ottoman gunsmiths knew about European weapons, it also seems that they were the source of some key innovations, particularly in relation to a trigger mechanism known as the 'serpentine'. Turkish infantry units were regularly equipped with matchlock muskets by 1500, and used them during successful campaigns in Iran and Egypt.

In the whole of this region, the introduction of firearms presented an awkward challenge to long-established military traditions associated with horses, which had been crucial in Arab armies, and even more prominently so among the nomadic ancestors of the Turks. Horsemanship still played an important part in upper-class life, and the most prestigious unit in any army was the cavalry. The troops responsible for artillery or hand-guns, by contrast, were often drawn from the lower strata of society, and could even be slaves. In the fourteenth century, as described in chapter 3, guns were sometimes used before they were really effective, because they symbolized power and could intimidate enemies by their noise. Now, ironically, when guns had become powerful weapons, their use was sometimes impeded because the mounted warrior was a more powerful symbol of aristocratic valour and manly virtue.

The Turks were more successful than other Islamic states in overcoming this difficulty. They nurtured a special infantry force with its own separate traditions and symbols, and a very firm discipline. These troops were known as the Janissaries, and had originally functioned mainly as archers. Their separateness from the Islamic ruling class was emphasized by the fact that they were recruited at an early age from the sons of Christians living within Ottoman domains. Legally they were slaves, but their status and career prospects were so good that many families were happy for their sons to be subjected to the rigorous discipline and training in the Islamic faith which the Janissaries' life entailed. It was these troops, then, who handled artillery or formed units equipped with muskets, and who made sixteenth-century Turkish armies so formidable.

A crucial encounter with the revived Persian Empire occurred in 1514. Although the Persians possessed some artillery, most of it was too far away to be used, so their army consisted mainly of horsemen. It faced Turkish guns and was decimated. The armies of Mamluk Egypt were similarly unable to match Turkish fire-power in 1516 and 1517, with the result that Egypt and Syria were incorporated into the Ottoman Empire.

Gunpowder and society

Despite losing some territory in these wars, the Persian Empire remained independent, but it had to fight the Turks several times more before a peace was agreed in 1555. After that, and particularly after Shah Abbas came to the throne in 1587, Iran was the scene of a brilliant flowering of Persian culture. Moreover, there was a strong Persian influence on the arts in both Turkey and North India. During the invasion of 1514, the Turks had occupied Tabriz, the former Mongol capital of Iran, and it is said that a thousand artists and craftworkers from that city were compelled or encouraged to move to Istanbul. This strengthened existing links between the arts in the Ottoman Empire and Persia, and was of significance for some technologies also, notably textiles.

Persian influence in North India was associated with the foundation of the Mughal Empire in 1526, and developed impressively after the accession of the Emperor Akbar thirty years later. The Mughal emperors could trace their descent from Chingiz Khan, and they had appropriate military interests, but their language and culture were Persian, and they attracted many Persian craftworkers to Delhi and Agra.

Ottoman Turkey, Mughal India and Persia all enjoyed times of considerable power and prosperity in the sixteenth century. On land, if not at sea, Turkey was the most formidable military power in the world for much of the century, and in all three empires, firearms and other gunpowder weapons were an important factor in territorial expansion. Thus Marshall Hodgson, the distinguished American historian of Islam, referred to Turkey and Persia as 'gunpowder empires', extending the term also to Mughal India. His point was that artillery, hand-guns and sometimes rockets not only allowed these empires to expand, but also led to a greater centralization of government. This was because the acquisition and deployment of guns required more resources and better administration than local potentates could achieve. When warfare depended on horses, the owners of land and livestock provided governments with their most essential military resources and formed an aristocracy with considerable influence. Now that governments had to buy in copper and tin, control the manufacture of arms and raise infantry units trained to use hand-guns, the traditional upper class was less important.

The concept of a 'gunpowder empire' has been taken up by several historians,[3] and seems likely to have the same sort of influence as Wittfogel's notion of 'hydraulic civilizations'. Both ideas refer to aspects of technology whose use is supposed to have crucially influenced government institutions and social structures. But while Marshall Hodgson defined a 'gunpowder empire' with some care, and only applied the term to a limited group of Islamic states, others have used this phrase to describe all the major sixteenth-century empires whose expansion depended on guns. Despite the lack of precision involved, this widening of the basic concept can usefully emphasize the very considerable expansion of several empires in mainland Asia, including those of Russia and China.

However, the idea of a 'gunpowder empire' does not work so well for Japan (which will be discussed separately), nor for Europe, and is basically at fault in attributing too much to hardware and too little to institutions. Ottoman Turkey, Mughal India and Persia can validly be grouped together as gunpowder states, not only because of the way they used guns, but also because of similarities of culture and institutions, especially military institutions and their relationship to court life. Institutional development in other empires of the time did not give the same prominence to the military, and so should not be so readily characterized in terms of gunpowder.

For example, when W. H. McNeill classifies China as a gunpowder empire, he is thinking of the situation after 1644, when the Manchu or Qing (Ch'ing) dynasty came to power.[4] China then responded to western and Russian imperialism by using its powerful armies and its guns to consolidate a long-standing connection with Tibet and to take over much of Xinjiang and Mongolia. Chinese expansion continued until just before 1700, when the steadily extending frontier of Russia was encountered. A first treaty defining the common boundary was signed in 1727.

During most of the sixteenth and seventeenth centuries, however, China was totally unlike the Islamic gunpowder empires. Its ruling groups were not a military class, but a body of professional administrators whose leisure interests were oriented to land-ownership and literature. Thus, whilst a large army had to be maintained, these civilian officials were always concerned to ensure that it remained under their control. Their interest in the land and in agriculture made them keen to ensure that farmers were not too heavily taxed to pay for arms. They encouraged new crops such as maize and sweet potatoes from the Americas. But above all, they read very widely. In 1600, it was said that more books were printed each year in China than anywhere else in the world. Thus if we are to characterize empires according to the technology most used by their ruling groups, the reference in China should not be to gunpowder but to printing.

One final point to make about China and Russia in relation to the Islamic empires is that territorial expansion was possible at this time because of a power vacuum on the Central Asian steppes. The role of Mongol nomads had long been declining, perhaps because their population was reduced by plague, but more particularly because their military tactics continued to be centred on horses, and they never made effective use of guns. Thus the expanding empires of China, Persia and Russia took over sizeable chunks of the former Mongol empire, Russia taking the biggest area of all. Much of European Russia had, of course, itself been ruled by the Mongols, and this 'Tartar yoke' had been a heavy burden. Around 1480, the Muscovites had made themselves independent of the Mongol khanate in southern Russia, and had gradually consolidated their control over the territories on the European side of the Urals. Expansion to the east began seriously in the 1580s, and proceeded so rapidly that an administration for Siberia was set up in 1637.

Gun manufacture

Some historians have implied that Indian mastery of gunpowder technology was minimal, pointing out that many Turkish and European mercenaries were employed by Indian rulers either to make guns or to operate them, and that large numbers of guns were imported. Certainly, imports were fairly common and there were important transfers of technology from the Turks. However, a large proportion of the guns used in India were manufactured there and local innovations did occur. When Akbar came to the Mughal throne in 1556, he took a personal interest in the development of his arsenals and in techniques for casting cannon. One distinctive feature of Indian gun manufacture from this time onwards was that many cannon were cast in brass rather than bronze. In Europe, brass was made by heating copper with a mineral named calamine, which was an ore of zinc. Indian metal-smelters, by contrast, knew how to extract metallic zinc from local ores, and could alloy it with copper directly. This allowed brass to be made more cheaply than was possible in Europe, and it proved to be a good material for the manufacture of small cannon.

Some guns were also made of wrought iron, by shaping several lengths of this material so that they would fit together to make the tubular barrel. These were then hooped round with wrought-iron rings to hold them tightly in place. This technique was first used in Europe, and then by Turkish gun-makers, from whom it passed to India. There it was employed to make some enormous guns, usually for mounting permanently on the walls of major forts. They were too heavy to manoeuvre easily for aiming, and took so long to reload that they were of very little military use. But guns fulfilled other purposes than mere utility as weapons. They were also status symbols, they expressed dreams of power and they could represent the more formidable of manly virtues. This is clear from the way in which many a brass cannon made in India was decorated with a tiger's jaws, symbolic of ferocity and power (figure 17).

Guns were never made of cast iron in India, but this material had been used in China from the 1350s. There was a danger with cast iron that excessively brittle barrels would be produced, which could burst when

Figure 17   An Indian 'tiger gun' – the brass barrel of a cannon cast in India during the eighteenth century.
(Illustration by Hazel Cotterell, based on an example at the Tower of London.)

the gun was fired. However, Chinese foundry-men had long experience with specialist applications of the metal, as when it was used for making bells (see chapter 1), and they could apparently produce satisfactory guns. In Europe, as blast furnaces developed slowly in the fifteenth century, cast iron was at first used only to make cannon balls. Some tentative experiments in gun-casting were tried in England in 1490, but quantity production was not attempted until 1541.

The relative cheapness of cast-iron guns was of great interest to the Dutch, because they needed many cannons for the ships of their large navy, and because of their wars with Spain. Thus the Dutch were responsible for the transfer of this technology to countries with which they traded, notably Sweden in the 1570s and Russia in the 1630s. The Russian foundry was at Tula, south of Moscow, and at first the guns made there were of poor quality. In due course, however, the manufacture of cast-iron cannon became a major industry in Russia. It expanded very rapidly after the development of ironworking in the Urals during the reign of Peter the Great. By 1715, some 13,000 cast-iron cannon had been produced in his foundries there.

With regard to hand-guns, some of the most interesting developments were to be found in Turkey, Iran and India. In many respects, there was a single culture of technology overlapping these three gunpowder empires, with Turkish or Persian craftworkers often working in India. Throughout the area, similar techniques for producing iron and steel were practised, with regional variations depending on the quality of the iron ore and of the availability of fuel, as well as on specialized local methods. Some of these techniques were very old, and as early as AD 600 steel blades made in Iran were being exported to China, possibly with some transfer of technology in that direction also. There were important steel-making centres in Iran, Iraq and Syria, but some of the highest-quality steel, known as *wootz*, was made in India.[5] Everywhere at this time (and until the 1850s) steel could only be produced in small quantities and at the cost of a great deal of fuel and many hours of labour. This limited its use very largely to making blades for knives and swords. Of the latter, the most highly prized were swords of 'Damascus steel', originally made in Syria, often from imported Indian steel.

The Damascus blade was characterized not only by its functional qualities but also by the etching of the surface of the steel with acid (often basically lime juice). This revealed an attractive pattern of irregular wavy lines associated with variations in composition of the metal. In Indonesia, similar techniques were employed in making a kind of dagger known as the *keris* or *kris*. It is said that Javanese smiths required over fifty days of repeatedly forging and reforging the steel to make one blade. Not only was the finished surface etched to reveal patterns caused by the heterogeneous nature of the steel, but the effect was enhanced by using nickel-bearing iron in making the steel. Whilst some of this iron was obtained from meteorites, nickel ores were probably worked on the island of Sulawesi (Celebes), where blades were made from about 1520. This could be the only pre-industrial nickel-smelting anywhere.[6]

From the sixteenth century onwards, steel produced in the same way as for Damascus blades was used in Turkey for making musket barrels. The steel was first forged into a long strip. This was then coiled into a spiral, using a simple machine with a heavy fly-wheel to twist the spiral quickly while the red-hot metal retained its heat. When the edges of the strip were firmly in contact (figure 18), they were welded together so that the coil (spiral) became a uniform tube. Finally the complicated barrel was etched with acid to show the damascene pattern running along the gun with the same alignment as the welded seam.

The first muskets to be made this way were probably produced within the Ottoman Empire, but by 1595 the technique was also being used in Akbar's arsenals in Mughal India. The barrels were stronger and less liable to burst than European barrels with longitudinal seams. Thus while Turks and Indians appreciated European muskets for their mechanical parts (and imported many), Europeans prized Turkish barrels, and the best European gun-makers sometimes used the Turkish barrels as the

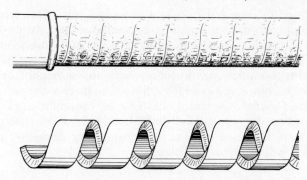

Figure 18   Coiling steel strip to make the barrel of a musket.
The bottom illustration shows a partly made coil, while at the top is a short
length of the finished barrel showing the spiralling damascene pattern exposed
by etching.

basis for guns otherwise of their own manufacture.[7] This is a very clear
example of technological dialogue.

Until a metallurgist named P. A. Anossoff studied steel-making in Iran
in the 1820s and began to manufacture steel of similar quality at a
Russian ironworks, European technologists were completely baffled by
the high quality of Turkish musket barrels, Damascus swords, and Indian
wootz steel. Nothing comparable could be made in any of the western
countries. One factor was the quality of the ores used by Indian steel-
makers. More significant, though, was their distinctive technique. In
Europe, steel for sword blades was made by heating wrought-iron bars
with charcoal. This was the 'cementation' process in which the iron
absorbed carbon very slowly at its surface *without melting*. Whilst some
steel was made in Iran by a similar method, most Asian steel-making
involved partial melting of the metal.

In 1740, a steel-making process which also involved melting the metal
was introduced at Sheffield (in England) by Benjamin Huntsman. A
comparable process was soon being used in France. The 'crucible steel'
so produced was of the high quality necessary for making tools for lathes.
However, the pattern to be seen on some Asian blades was never
obtained, and this pattern, together with the high quality associated with
it, still puzzled western steel-makers. Thus even in the 1790s Indian
wootz steel was the subject of investigation in Sheffield, where it was
used to make specimen blades of a quality which could not be replicated
by other means.[8]

Furnaces capable of producing cast iron (often on a very small scale) were operated in Iran and Iraq as well as in China. One steel-making process employed in both regions has been described as 'co-fusion'. It consisted of melting small quantities of cast iron and wrought iron together. The point here, if we look at it in modern terms, is that cast iron contains too much carbon for strength and wrought iron contains too little for hardness. If a carefully adjusted mixture of the two is melted, the combined carbon content would give an optimum combination of strength and hardness in the finished blade. Several other steel-making processes were also used in Asia, including a very distinctive process in Japan which will be discussed below.

Trade, textiles and ceramics

C. G. F. Simkin has remarked that the handicraft industries of India and China were the twin pillars of commerce in Asia. Cotton textiles were the most important of India's exports, but here again there was a sharing of technology with Iran and Turkey. In addition, there were smaller handicraft industries in Sri Lanka, Burma, Thailand and Java which drew exports of raw materials from India (raw cotton and dyestuffs), from China (silk) and from Japan (copper). Much sea-borne trade was, by the later sixteenth century, carried in European ships, but most of this was trade between one Asian country and another. Exports to Europe were a small proportion of the whole. Russia also carried on an extensive trade with its Asian neighbours. In the seventeenth century, there were Indian merchants in Moscow and Astrakhan, and caravans plied between Moscow and China.

In order to understand manufacturing in the gunpowder empires, we need to recall an earlier point about how governments became stronger and more centralized as a result of their control of gunpowder weapons. In many cases, government control extended to some aspects of manufacturing also. The aim was not to promote trade but to supervise production of quality textiles and furniture in support of a magnificent style of court life. This led to a form of industry characterized by royal factories or *karkhanas* in Mughal India, and by 'imperial workshops' in Ottoman Turkey.[9] In such places, the system of government supervision first used in manufacture of weapons for the army was applied to other manufactures, such as silk. At one time there were 4,000 silk workers in *karkhanas* in Delhi, whilst in the Ottoman Empire the court gave work to a similar number, some in the imperial workshops but many in the private sector. Many of the latter were based at Bursa, a centre for the silk trade at the end of the overland routes from China and Iran. Whilst some silk was produced there by raising silk worms locally, much raw silk was also imported from Iran. Indeed, Italian merchants came to Bursa to buy the raw product for manufacture into thread and fabric at Lucca or Bologna.

There are few technical details of the looms used in the sixteenth century, but much can be inferred from contemporary looms in China and from looms worked more recently in Turkey and Iran. In all three countries, much silk was woven with elaborate patterns on large draw looms which had to be operated by two people. However, there were also many smaller looms such as the Persian type illustrated (figure 19) for local production of silks and cottons of more ordinary kinds. A feature of most looms was the use of treadles to raise and lower the heddles, leaving the weaver's hands free to pass the shuttle to and fro.

In comparison with the large scale of operations in the 'royal factories', the production of cotton textiles for export from India was mostly carried on with very simple equipment in numerous scattered villages. Merchants advanced the necessary materials and then, despite poor working conditions, quality production was achieved through a high degree of

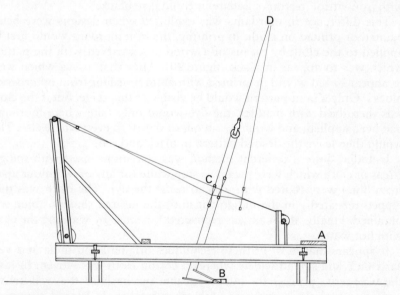

Figure 19 Outline drawing of a small Iranian loom for weaving narrow fabrics in silk or cotton.

There is a seat for the weaver at A, and she operates the treadles (B) with her feet. The heddles (C) make a passage between the warp threads through which the shuttle is passed and are suspended from the ceiling of the workroom (D). For a detailed discussion, see Wulff.

specialization. Many skills were closely associated with particular castes, and it used to be said that this had the effect of tying workers to processes and inhibiting technological change. Historians have recently argued against this view, but the impression remains that in India, Turkey and other Asian empires, craftworkers often found themselves employed in rather tightly restricted conditions.

Whatever might be said about the effect of this on innovation, there were also important migrations of workers, often at royal behest, as when, in the late sixteenth century, Turks and Persians were recruited to work in India. This certainly led to important developments in technology. Taking cotton textiles and methods of dyeing as an example,[10] although India was the source of many important processes, techniques had developed differently in other countries. Two dyes were of particular importance: madder and indigo. Both were obtained from plants, but they had very different characteristics. In particular, madder could only be used successfully if the cloth were first treated with a mordant solution. This could contain any of several mineral salts, the choice depending on the colour required. With an alum mordant madder produced red, but with iron vitriol (ferrous sulphate) it could give black.

This difference in mordants was exploited when designs were being painted or printed on cloth. In printing, the iron mordant would first be applied to the cloth by means of a wood block carved with the pattern which was to appear in black (figure 20). After that, areas which were to appear in red would be printed with alum mordant from other wood blocks. Only a faint pattern would be visible at this stage, but if the cloth was then dyed with madder, the dye would only take where mordants had been applied, and would wash out of the cloth everywhere else. This would then leave the desired pattern in black and red.

In India, quite a different method was sometimes used with indigo. Areas of cloth which were meant to stay white (or any other colour apart from blue) were coated with wax to *resist* the dye. The cloth was then dipped repeatedly in the indigo vat until the desired shade of blue was obtained. Finally, the wax was recovered for reuse by washing the cloth with hot water.

It appears that most of these techniques originated in India at a very early date and spread widely wherever cotton cloth was woven. In Java, the wax-resist process for indigo became the basis of the well-known batik textile designs, executed with other colours. In Islamic regions of West Africa, where cotton textiles were also made, indigo dyeing was carried out in great vats sunk into the ground. In Iran, the printing of cloth developed particularly strongly, and during the Mughal period, from about 1550, Persian craftworkers transferred their techniques back to India, introducing them into new areas where cloth had previously

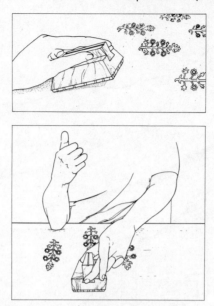

Figure 20   Two steps in printing cloth with a wood block.
First the block is carefully placed in position (top), and then it is struck sharply with the right hand to transfer the mordant solution to the cloth. The illustration shows the outline of a flower pattern being printed, with no other part of the design present, which suggests that an iron mordant is being used so as to reproduce the outline in black when the cloth is dyed with madder.
(Illustration by Hazel Cotterell)

been hand-painted rather than printed. Making wooden blocks for printing was itself a separate trade involving very fine work with bow-drills and chisels. By 1700 if not earlier, the centre of this trade in India was Ahmedabad, which was well placed to serve the textile printing centres of Gujarat. Meanwhile, Turkish craftworkers had devoted themselves to bringing out the red colour of madder more brilliantly, and had arrived at an elaborate process involving sixteen major stages, with oil and tannic acid employed for initial treatment of the cloth as well as the alum mordant. This 'Turkey red' process also filtered back to India.

Technological innovation in these handicraft industries was the result of careful observation and painstaking trials, and led to very small improvements over a long period. In these circumstances, many variants of basically similar techniques could co-exist, so that when an early Mughal medical book discussed vegetable dyes, the author was able to list 77 different processes for producing 45 shades or colours. In steel-

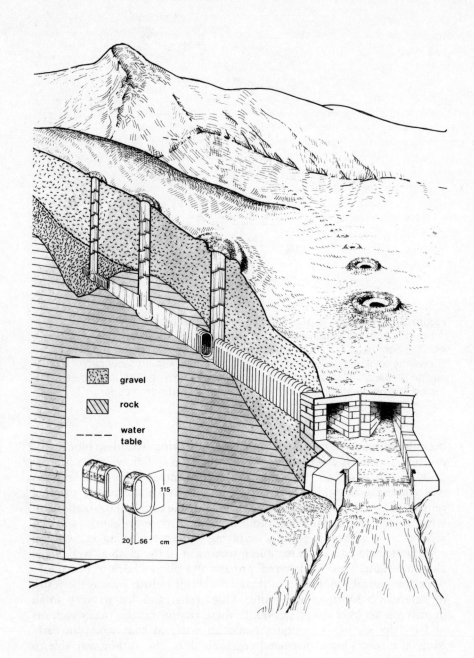

gravel

rock

water table

115

20 | 56 | cm

making and other technologies also, the accumulation of empirical knowledge was similarly expressed in many detailed variants in techniques and products.

It is tempting to suggest that these developments were symptomatic of conditions in which radical innovation had ceased and only detailed improvements in techniques were occurring. However, this is far from the truth. Technical progress in India had been greatly stimulated during the sixteenth century by the appearance in the region of European ships and weapons. In a remarkably short time, Indian arsenals were producing guns of good quality. The influx of Persian textile workers at the same time may well have reinvigorated some aspects of the traditional cotton industry in India. Its export performance in the seventeenth and early eighteenth century was very impressive indeed.

There were also significant developments in the Persian Empire, where Shah Abbas not only established 'royal factories' in the vicinity of Isfahan, but also promoted the pottery and glass industries with an eye to their commercial and practical importance. Fine pottery had been made in Iran for centuries. Tiles were manufactured as a facing material for buildings, either for use as complete tiles, or for cutting up to make mosaics. More utilitarian ceramic products were earthenware pipes for irrigation systems, and heavy earthenware rings for lining qanat tunnels where they passed through soft ground (figure 21). The importance of the latter can be appreciated from the estimate that more than half of all water used in Iran for irrigation or urban supply came from qanats. There were even some underground water mills constructed to exploit the flow of water in the tunnels.[11]

Because of Iran's position on the old trade route to China, it was inevitable that imports of Chinese porcelain would attract attention. Over the years, good imitations of some Chinese styles were produced, though not real porcelain. But here, as in other fields, there was dialogue

---

Figure 21   Two qanat tunnels in a typical landscape where rainwater running off mountains replenishes groundwater beneath the sand and gravel of the lower slopes. (Water-bearing gravel is shown darker than the dry gravel layer above.)

The ground penetrated by one of the qanats is shown in cut-away section to demonstrate how the tunnel gives access to water trapped by impervious rock. The section also shows how the tunnel is lined with earthenware rings when passing through sand or gravel layers which might otherwise cave in. Qanat tunnels were excavated from the bottoms of a series of well shafts, with the 'mother well' at the upstream end dug first to prove the existence of water. The tunnels were sometimes several kilometres long with their route marked on the surface by the line of well shafts, as on the right of the picture.

(Illustration by Hazel Cotterell)

as well as imitation. Cobalt blue glazes were a Persian innovation, and from 1360 onwards materials for making this glaze were being exported to China. In the 1580s, Shah Abbas was especially keen to strengthen the pottery industry, and given the long-standing connection with China, it is understandable that he invited three hundred Chinese potters to Iran to instruct local craftworkers. His commercial acumen was shown by the way in which he also encouraged imports of porcelain from China, much of it for re-export to Europe. This allowed his merchants to bypass European shipping in the Indian Ocean and secured profits which would otherwise have gone to the westerners.

Japanese muskets[12]

Whilst Chinese trade with the Persian Empire prospered, cuts in all kinds of shipping ordered by a Ming government in the fifteenth century had left China's sea-borne trade to Japanese vessels, often operated illicitly by Chinese merchants. When the Portuguese began to trade in the area during the 1530s, they rapidly became involved in this illegal traffic between China and Japan. The first to visit Japan were three who were travelling on a Chinese-owned boat which was blown off course in 1542 or 1543. The people they met in Japan were immediately interested in their unusual weapons. The Portuguese visitors were asked to demonstrate what were, in fact, matchlock muskets, and then it was requested that local craftworkers should be allowed to copy them.

Reports of this adventure encouraged the Portuguese to send one of their own ships to Japan a year or two later. Then the missionary, Francis Xavier, arrived in 1549 bringing a clock and some western books on astronomy, which attracted much interest. By the end of the century, Japanese craftworkers were regularly making clocks in small numbers. At first they copied imported examples, but then invented a double escapement designed to register the unequal hours of day and night which were a feature of time measurement in Japan.

The manufacture of both clocks and muskets developed rapidly because Japan already had many artisans skilled in metalwork. As with so much of Japan's earliest technology, many of their basic methods had originated in China. However, there was a distinctive steel-making technique (the 'Tatara process') which appears to have evolved in Japan to cope with iron-bearing sands and gravels found there.[13]

These raw materials were smelted in a low rectangular furnace. After being charged with 'iron sand' and charcoal, the furnace was kept burning for three days, with a continuous air-blast from manually operated bellows. When the process was judged to be complete, the furnace was allowed to cool, its walls were broken down, and a large irregular mass of iron was revealed. This was of uneven quality because only the centre of the furnace was hot enough to melt the metal completely. When it was smashed into fragments, workers could distinguish different types of iron by eye and sort them for forging or casting, as appropriate. A typical result might be 1.7 tons of pig iron, easily distinguished by its brittleness, 0.8 tons of steel and a similar amount of soft iron capable of being forged into wrought-iron bars. In 1600, and for the next three centuries, most iron and steel produced in Japan was made by this process.

Much of the steel output of Japan was used to make swords, which became famous for their high quality. They were comparable to the best 'Damascus' blades, and some were exported. Thus there were many artisans with appropriate skills for making guns, and it has to be added there were many people keen to set up gun-making workshops. This was because Japan had for a long time been in a state of civil unrest, with local barons maintaining private armies which were quite often involved in fighting.

One of the most important of these warlords was Nobunaga, who had learned the use of firearms as a young man in the 1540s, and by the 1570s controlled the manufacture of guns in central Japan. He encouraged trade with the Portuguese so that lead for bullets and saltpetre for gunpowder could be imported, and he also pioneered the manufacture of cannon, melting down bells taken from temples to obtain the necessary bronze. He had thoughts about the use of cannon on ships as well, and experimented with a warship whose hull was armour-plated with iron.

Guns had been used in warfare in Japan several times since the 1540s, but a battle fought in 1575 was the first time they had decided the outcome of a conflict. Nobunaga had used 3,000 infantry soldiers armed with muskets, and had worked out tactics for using the guns to advantage. The muskets themselves were still very crude, and did not work reliably in wet weather. To remedy this, Japanese gun-makers invented a box-shaped matchlock protector to keep rain off the match and powder. An interesting comment on the quality of these weapons is provided by a Chinese book written in 1598 which discussed muskets made in Japan as compared with Turkish and Portuguese examples – hardly any had been made in China by then. The Turkish weapons were preferred, one problem with the Japanese guns being the position of the touch-hole, too close to the user's eyes when taking aim, and hence a possible hazard.

At Nobunaga's death in 1582, his ruthless military campaigns and his mastery of tactics based on firearms had brought fighting in Japan to a halt, which gave an opportunity for more moderate leaders to encourage an agreement for national unity. Some progress was made, but in 1600 internal dissension led to more fighting. Once again, guns played a decisive part in the outcome, and at the siege of Osaka Castle artillery provided by the British and Dutch was used to good effect, Japanese cannon being too small and light. A settlement inaugurating a 'great peace' was finally achieved in 1602–3 when the first of the Tokugawa family to become *shogun* took office – that is, he became effectively viceroy, exercising political power in the name of the emperor.

The Tokugawa shoguns adopted a wide variety of stratagems to prevent further internal wars, such as insisting that every baron should maintain family members at court continually as hostages for their own loyalty. The shoguns were also very wary of potentially disruptive influences such as missionaries, who were now banned, and European traders, who were eventually excluded (except for the Dutch, and they were allowed only a small trading post on an island in Nagasaki harbour). Meanwhile, the economy prospered. The area of land under cultivation doubled as marshes were reclaimed. New crops were introduced, including tobacco. Rice production had increased by about 30 per cent before 1700, with improved irrigation from canals, and hydraulic technology transferred from China, including treadle-operated water-lifting devices. Textile output also expanded, with spinning and weaving mostly carried on as a household industry. More land was devoted to the cultivation of crops such as cotton and indigo, with mulberry trees being planted to support the expansion of silk-worm raising. Mining increased, and much copper and silver was exported for use as coinage in China. Copper was traded to India in Dutch ships for making bronze or brass cannon, and the Dutch also took considerable amounts back to Holland for gun-casting.

To the extent that the introduction of muskets played a large part in ending the 'baronial wars' in Japan and in enabling the Tokugawa shoguns to establish a strong national government, Japan conforms neatly to the pattern shown by other 'gunpowder empires'. However, Japan differs from the others in several respects, most notably its less centralized form of government. National unity did not mean a national bureaucracy in the Chinese manner, and the nobility still had considerable power in their own provinces.

It should also be stressed that although European merchants (except the Dutch) could no longer trade with Japan, there was a good deal of sea-borne trade, especially with Korea and China, and this sustained a prosperous merchant class. Moreover, the success of this commercial activity was not compromised, as sometimes in Turkey or Mughal India, by heavy taxation to support military campaigns, nor by the removal of skilled craftworkers to non-commercial royal factories. Also, in contrast with the gunpowder empires, Tokugawa Japan held its military institutions within strict limits, and it is striking that the manufacture of guns was one branch of technology which, after 1650, stagnated or even declined.

6 Concepts in technology, 1550–1750

Developments in printing[1]

Printing was a technique of crucial importance for many other branches of technology because of its role in spreading information and ideas. But while all the great kingdoms and empires of Asia were manufacturing guns by 1600, printing had not yet been adopted in the Islamic countries. This might seem surprising, since it had been known in that region since 1294, when Chinese printers were employed in Iran to produce notes for an experimental issue of paper money.[2] In the sixteenth century, there was prosperity and the arts flourished, but still printing was not introduced. One reason was that the demand for books could be adequately met by hand copying. However, there was also opposition in conservative religious circles. It was regarded as an act of piety to make a new copy of the Qur'an (Koran) by hand, and calligraphy was seen as a fine art. In Turkey, many eminent people – even sultans and viziers – would copy the Qur'an, and a professional calligrapher might make 50 copies in a lifetime. It took a firm lead from a reforming vizier in the 1720s to introduce the first Turkish press and use it to print translations of western books.

As to India, a press was set up by the Portuguese in Goa, and a book was printed in the Tamil script in 1578, but aristocratic society showed no interest. Again, part of the reason was that all the books required by a small literate class were produced cheaply enough by numerous professional calligraphers. One might expect a different situation in Russia, but although printing was introduced in 1524 with the help of Danish craftworkers, it was used mainly by the Church to produce a very limited range of material.

The contrast between rapid adoption of guns and slow take-up of printing in these countries is highly significant. Guns served the interests of existing ruling groups, who were often military elites, and allowed them to consolidate their power and extend territorial boundaries. Printing was irrelevant to these goals. However, in China, as the previous chapter indicated, the situation was quite different. The imperial civil service was a literary elite, whose members were interested in the arts,

Confucian philosophy, and their country estates. Their demand for books was insatiable, and stimulated innovations such as colour printing. The first books with illustrations printed in colour were produced in China in the 1580s using a five-colour system. However, experiments in printing with two colours (red and black) had been made as early as 1340.

Korea and Japan used basically the same written language as China, even though speech was very different, and printed books were exported from China to both countries. There were well-established printing workshops in Korea, but in Japan wood-block printing was done only in Buddhist monasteries. About 1590, Jesuit missionaries introduced a western-style press with movable type. This helped stimulate the setting up of commercial printing shops independent of the monasteries. Some of the equipment and type was obtained from Korea after an invasion in 1592, and classics of Japanese secular literature which Buddhist printers had ignored were set in type for the first time.

By 1640, however, movable type of both Korean and western varieties had gone out of use. Japanese printers had reverted to wood blocks. Movable type did not suit the very free format of much popular literature. Of more importance than the printing methods used, however, was the fact that new books on technical subjects were being produced in Japan, including works on navigation (1618) and mathematics (in the 1620s). In a few of these books, the Chinese colour-printing method was used in a limited way to make diagrams clearer. The printing of art books and prints in colour became much more important in China and Japan after 1700.

Some good technical books were also printed in China in the sixteenth and seventeenth centuries, dealing with such subjects as firearms and iron-smelting, but they represent only a small part of the total of books produced. Some historians[3] have argued that the Confucian ruling class lacked interest in mathematics or physical phenomena, and such subjects certainly attracted less attention than in Japan. There was perhaps a tendency, basically realistic, to see the solution of practical problems in terms of good organization and effective government. Technology was left to craftworkers and entrepreneurs.

This attitude must have seemed fully justified in that China had a strong economy in which agricultural output was increasing and many industries flourished (including textiles, ceramics, iron-smelting and, of course, printing). However, what has prompted historians to talk about 'stagnation' in Chinese science and technology is that this was the period of the 'scientific revolution' in Europe, and it was a period, too, when European guns, ships and military techniques were developing fast. A Jesuit mission in Beijing kept the Chinese authorities informed on these matters and even provided some gun-making expertise during the crisis which led to the downfall of the Ming dynasty in the 1640s, but no comparable scientific movement developed in China.

To evaluate the issues this raises, we need to be clear about ways in which the 'scientific revolution' may have contributed to the technological advantage which the West subsequently gained as compared with other cultures. The argument of this chapter is that what mattered was not any particular discovery or invention, but rather a series of new methods for handling technical information and formulating technological (and managerial) ideas by analysis based on measurement, tabulation of data, classification or subdivision, or even the use of drawings and physical models.

Blocked technical systems?

One western innovation whose adoption in other cultures has attracted specific comment from historians is the weight-driven clock. Whilst water clocks were still used in China and India, Japan was the only Asian country where clocks of the western type were being made in the seventeenth century, and where new designs were evolved. Lack of clock-making workshops elsewhere is sometimes cited as evidence that Asian technology was 'stagnating', with what one author describes as 'blocked technical systems'.[4]

Taken at face value, such talk is nonsense. There was continuing technological development in much of Asia, involving shipbuilding and textiles in India and the wide range of Chinese industries already mentioned. One can also point to individual inventions, such as colour printing, and to the introduction of new crops, such as maize and sweet potatoes. Over a longer period, culminating around 1650, new equipment was introduced into the Chinese textile industry to cope with an increased use of cotton in place of other fibres (including hemp and ramie).[5] Machines used with other fibres went out of use or were adapted to serve different purposes (notably the one illustrated in figures 9 and 10). Machines which took their place included cotton gins, a new three-spindle spinning wheel, and calendering rollers (used in large enterprises which specialized in finishing cloth). In one region where this activity was concentrated the number of people employed in the calendering industry grew from 7,000 around 1700 to 11,000 in 1730, and other branches of cotton manufactures expanded correspondingly. In the silk industry, elaborate draw looms for weaving complex patterns had become common during the Ming period, and were probably related to very similar looms used in Persia.

In these respects, then, China's technical system was clearly not 'blocked'. Invention and expansion were continuing, and responded effectively to pressures arising from resource scarcity (which encouraged innovation) and a growing labour force (which limited the use of some labour-saving machines).

However, in one respect it may be right to envisage 'blockages' which could limit development of technology in some cultures as compared with others, because with many techniques there is a limit to the improvements which can be made by craftworkers' methods. For example, it is now known that dye made from madder plants grown on calcium-rich soils gave a specially deep shade of red. Techniques were traditionally adjusted quite empirically to allow for differences in the quality of madder without any knowledge of calcium and its effects. However, empirical development along these lines could reach a stage where further progress was blocked by lack of theoretical understanding.

This kind of blockage could also occur in metalworking or the design of machines, and the way round it usually depended on analysing the process more clearly so as to conceptualize what was involved. To do this, it was often important to record data, perform a chemical analysis, or draw a mechanism on paper. The advantage gained was seen after the Turkish version of madder dyeing was introduced into France in the 1740s. A French chemist, Duhamel de Morceau, soon recognized the role of calcium (or lime) in bringing out the red colour,[6] and then had a firmer base for suggesting improvements.

With mechanical technologies, a comparable procedure for analysis and conceptualization often depended on measurement and scale drawing. The latter technique is said to have been 'invented' by the Italian architect Filippo Brunelleschi just before 1420. Certainly, architects and map-makers were increasingly making drawings to scale from about this time, and soon shipwrights such as the Englishman Matthew Baker were doing the same. Another technique used in ship design was the construction of scale models, and Galileo, the great Italian mathematician, wrote about the comparative strengths of models and full-size structures. This was in 1638, and indicated yet another new analytical approach. Galileo's theory had occasional use (for example, in the 1670s and 1790s) when tests were made on models of pipes, beams and bridges, and it was desired to apply the results to full-size structures.

Printing has been stressed here because the ready availability of books was a stimulus to the growth of habits of abstract thought, often analytical, in China and Japan as much as in the West. But while there were many seventeenth-century Europeans who applied such analysis to problems of map-making or mechanism – to wheels, levers and chemical substances – in China, people with the same inclinations applied them instead to the analysis of documentary evidence or linguistic problems.[7]

One symptom of this difference in interests is attitudes to clocks. In the West they were seen as of great significance in showing complex motions of the planets reduced to a model with simple parts. However, there is a difference between *models* as straightforward representations of material objects, and *symbols* as signs of profounder realities. Seen as a model, the clock encouraged the construction of other mechanical models, some of them merely clockwork toys, made to entertain. Seen as a symbol, the clock represented the growing belief that a form of *mechanical order* permeated the whole universe. This encouraged people to conceptualize natural processes in terms of mathematical relationships known to be applicable to machines such as clocks.

Intellectual interest in astronomical clocks in China and the Islamic world during the eleventh and twelfth centuries was mentioned in chapter 2. This can be taken as evidence that some conceptualization of 'mechanical order' was already being attempted. In both civilizations there were important developments in mathematics at the same time, notably trigonometry and algebra in the Islamic world. But by 1601, when Matteo Ricci, a Jesuit priest, presented a European clock to the Chinese emperor, the literary and agricultural emphasis of Chinese culture was so strong that, as Ricci noted, nobody with real ability took up mathematics, and no new work was being done. The great clocks made before 1100 no longer existed, and the thinking behind them had been forgotten. So the European clocks which were now being introduced were not recognized as symbols associated with cosmological ideas. They were collected along with other mechanical novelties as 'intricate oddities'. The Chinese writer who used this phrase added that they were 'attractive to the senses', but 'they fulfil no basic need.'[8]

It is significant that astronomy was an important stimulus for this kind of work, and that astronomical observatories were the most characteristic institution for dissemination of ideas about mathematics, clocks and some types of technical drawings (especially scale maps). In earlier chapters we noted the existence of observatories in India, the Islamic world and China. Discussion of why clockwork or mathematics failed to develop as vigorously in Asia as in Europe is linked to the question of why most of these observatories were destroyed or abandoned during the period of nomadic invasions. During the sixteenth century, and probably during the reign of Akbar, the ancient observatory at Benares was restored. Later, other observatories were constructed or re-equipped in India and in China under the influence of European astronomy.

In the Islamic world, there was a small but significant revival of interest in the philosophical aspects of clocks, as well as in drawing and mathematics, when an observatory was set up at Istanbul between 1575 and 1577. Not only was it equipped with astronomical instruments, but it was organized for geographical work as well, with a globe and maps.[9] These last probably included work by Piri Reis, a Turkish naval officer famous for maps of Mediterranean ports and of the Atlantic, drawn before 1521. The man responsible for the observatory, Taqi-al-Din, was aware of earlier Islamic writers on mechanics and produced his own book on machines, influenced by the thirteenth-century work of al-Jazari (chapter 2). In 1565, Taqi also wrote about clock mechanisms, discussing all the latest western types. He built a clock for the observatory, but noted the low price of clocks imported from Holland and Germany, and doubted whether Turkish craftworkers could compete.

The Istanbul observatory did not last long, but was closed as a result of pressure from much the same conservative elements as resisted the introduction of printing. As a result, the cartographic works of Piri Reis and the technical books of Taqi-al-Din had few successors. Thus the empire which had played a leading role in the development of guns failed to nurture the roots from which fresh technological thinking might have grown. What mattered for the future was not just hardware, but also the kinds of thinking expressed in drawings, tables of figures and printed books.

From these points of view, the most significant developments in Asia were the technical books published in Japan during the seventeenth and eighteenth centuries, a handful of Chinese scientific works, and very occasional episodes in India such as the use of models in the design of the Taj Mahal in the 1630s, and the systematic use of scale drawings by some shipbuilders by the end of the eighteenth century. But such examples are few and isolated. The great preponderance of new technological potential generated by increased ability to conceptualize technical problems was accruing in the West.

Concepts of organization

Conventional interpretations of the change in outlook which occurred in Europe between the fifteenth and the seventeenth centuries tend to obscure its relevance to technology by reference only to a 'scientific' revolution. This is often narrowly defined, without recognition of new ways of thinking about practical problems which were emerging, relevant to management, organization and technology.

Yet it is generally acknowledged that the writings of the English lawyer, Francis Bacon, captured the spirit of the new approach in important respects. And Bacon not only wrote about how science should be studied in terms of collecting, classifying and analysing facts. He also put forward ideas about how scientific work should be organized, and how it could benefit from collective effort with an implicit division of labour. Thus the questions to ask about the influence of the scientific revolution on technology do not simply concern the application of scientific discoveries in newly invented techniques. Examples of this are few, the most important being the application of discoveries about atmospheric pressure by the inventors of the steam engine. More important are the questions indicated by Bacon's work about how problems should be analysed and operations organized.

For example, in the seventeenth century, the habit of thinking about processes in terms of machines spread from ideas about clocks and the universe to the study of the human body (the heart considered as a pump), and then to the study of thinking processes (with the invention of a calculating machine by Pascal), and finally to ideas about human behaviour and organization. As Carolyn Merchant puts it in her reinterpretation of the scientific revolution, 'the machine . . . functioned symbolically as an image of the power of technology.' Whilst the clock served as 'a symbol of cosmic order', she cites a seventeenth-century European context with which heavy dockside cranes operated by man-powered treadmills 'symbolize the role of the machine in organizing human life'.[10]

In one instance after another in Europe at this time, we find people analysing how individual 'machines' were operated, including muskets, surveying equipment and spinning wheels. In each case, the analysis included the motions of the operators' arms and fingers, either with a view to redesigning the machine, or else with the idea of reorganizing the task so that the operator worked faster. Such reorganization could involve training the operator in more efficient ways, or dividing the task between several people.

As an example, we can consider the work of Maurice of Nassau, who was Captain-General of Holland around 1600, when the Dutch were fighting Spanish domination of the Low Countries. He had been a student of mathematics and classics, and had taken particular interest in Roman military procedure. In commanding the armies of Holland, he made three innovations based on Roman precedent.[11] One was to insist that his troops should carry spades and dig themselves into good defensive positions. Another was to divide his army into smaller units than had been customary so that a single voice could issue commands that would be heard by the whole unit, and so control its movements.

The third of Maurice's reforms was systematic drill, both in marching and in the use of weapons. Maurice studied the rather complicated actions required to load and fire matchlock muskets. He analysed them as a sequence of 42 individual moves and, after illustrating each move with a drawing, gave it a specific word of command. Soldiers could then be taught to make the moves in unison in response to shouted commands. He found that after practising the sequence of movements many times, soldiers made fewer mistakes and could fire more rapidly. Moreover, performing the movements in unison made it possible to fire devastating volleys.

The last of these innovations indicates an approach which was to become very characteristic of European activities in many fields during the next two centuries. The analysis of an operation such as loading a gun had parallels in William Petty's approach when making a map of Ireland in the 1650s. In this instance, it was the use of surveying instruments and the recording of information obtained from them which had to be analysed. Petty reduced the whole procedure to a large number of small operations which could be allocated as separate tasks to unskilled soldiers. This is often quoted as a pioneer application of the systematic division of labour.

In the next century, the same sort of analytical thinking was applied to spinning with a wheel to produce cotton yarn. It was observed that women operating spinning wheels performed an important function with their fingers as they fed a roughly twisted sliver of cotton fibres into the process, and a second function using movements of their arms. The aim in analysing these actions was not to make the operator more efficient but to replace her by a machine. It was recognized by Lewis Paul around 1738 that one or two pairs of rollers could be used to feed the sliver into the machine in the same way as the spinner used her fingers. Then various ways of mechanizing the movements of the spinners' arms were tried, culminating in the invention of two separate but equally successful machines by Arkwright and Hargreaves in the 1760s.

Arms drill, division of labour and the design of industrial machines all required a similar approach, therefore, consisting of the analysis of complex motions made by arms, hands and fingers into many simpler component motions. This approach could allow the performance of a task to be reorganized so that the people who carried it out worked together in a more unified way. Every man in Maurice's well-drilled armies fired his gun simultaneously so that the firing came in volleys rather than in a randomly timed smattering of bullets. At this time, no muskets could be fired very accurately. Rather than letting soldiers spend time in taking aim, therefore, effectiveness was sought by collective fire repeated at frequent intervals.

It would be wrong to suggest that many people were consciously aware of new principles of organization, although readers of Bacon's works may have been. It was perhaps more a question of an intuitive reaction to an environment in which machines were more often seen and discussed, and were even used as analogies for social or biological activity.

One summary of how German historians see connections between military organizations and mechanization comments on the notion of a cavalry unit as a combat 'machine'. The latter word came to be frequently used in relation to military and other organizations, and of all the machines people thought of, the clock carried the greatest symbolism. It had become a *model* of the good organization, and in addition, was being used as a *tool* for the regulation of organizations, as working hours were increasingly defined by the clock. Even an army's drill in marching or firing might be done to time.

Significantly, the German historians have seen the new approach to army training as analogous to the division of labour, and as a strong influence on the culture of Europe, especially from the 'decisive' seventeenth century onwards. The increased use of muskets from before 1600 in Holland, following a transition from pikes and halberds (in England from longbows), was comparable to the change in industrial production 'from a craft to an assembly line mode'.[12] It can also be argued that the new form of 'capitalistic production' which was developing in European factories by 1800 had been manifest earlier in European armies. The development of military organization thus influenced the later development of civilian industry.

Factories and plantations

The idea that military organization influenced the development of factory production is a very significant suggestion, but it can only be part of the story. The sociologist and historian Peter Worsley has identified another source of ideas for the organization of production as plantation agriculture.[13] He traces this to Portuguese sugar plantations on the island of Madeira from about 1420, where slave labour was used (with some white slaves at first, and then African slaves from 1450). Long before 1600, a carefully planned division of labour with strict work discipline had evolved. Worsley argues that such plantations were 'a model or prototype for the later organization ... of the manufacturing units named "factories" ... which employed wage-labourers under conditions of such intensive labour and such loss of control over their work that the workers called it "wage slavery"'.

It should be specifically noted that it was the *organization* of the sugar plantations which was novel at this time. The cultivation methods and cane-processing technology used by Europeans in Madeira and later on Caribbean islands had been acquired from the Islamic world, and from Sicily. Morocco had an important sugar industry during the fifteenth century. The north Moroccan town of Ceuta was invaded by the

Portuguese in 1415, just a few years before the colonization of Madeira began, and Morocco was probably one source of information concerning sugar technology, such as cane-crushing mills. The plantations on Madeira proved to be highly lucrative, and exports to Europe expanded fast. By 1493 there were 80 'factory managers' responsible for sugar production on the island.

When sugar plantations developed in the Caribbean region, technology derived from Islamic sources was again used. Then, in the 1690s, there was another transfer of technology from Africa to the New World. The area concerned was in South Carolina, and the crop was rice. English settlers had grown rice at an earlier date in Virginia, perhaps using the same methods as prevailed in Italy. However, an American historian, D. C. Littlefield, has suggested that a different kind of rice was found necessary for swampy land in South Carolina, and one contemporary source said that this was introduced in 1696 from Madagascar. While there may be truth in this, Littlefield argues convincingly that slaves working in this part of America came from a district in West Africa, near what is now Conakry, where rice had for a long time been grown under irrigation, with small reservoirs or ponds for water storage. Slaves from this area had skills useful to American farmers, especially the women, whose dexterity in transplanting rice (at '50 roots per minute') was noted by travellers in Africa. Similarity of technology between West Africa and South Carolina around 1700 extended to details of fertilization (with ashes) and the use of hollowed tree trunks to make sluices and drains.[14] Thus a transfer of technology from West Africa seems highly probable.

In this instance, as in so many others at this time, the distinctive contribution of Europeans to technology was a matter of organization for higher output through control of the work force. Meanwhile, practical techniques were adopted from wherever they were available.

Division of labour was one aspect of the new concepts of organization which were circulating in Europe, but travellers in Asia could not help noticing that division of manufactures into many component tasks was already a feature of handicraft industries there. It was said that in China, a porcelain plate or bowl would pass through the hands of 70 workers before it was finished, each one highly skilled in some detail of working clay, firing kilns, decoration or glazing. But the purpose of this division of labour was somewhat different from that which predominated in Europe. The aim was not so much to speed up production as to achieve very high quality by enabling each craftworker to acquire skill by specialization.

In India during the eighteenth century, British observers noted that cotton cloth would sometimes be worked on by four different people where only one person would do the job in Britain, and the remarkably fine muslin yarns produced in Bengal were the result of many detailed processes which could not be reproduced in the West at all. However, another part of the experience of European merchants in Asia was that they might order a consignment of cotton cloth printed with a specified pattern, but the time taken by Indian craftworkers in meeting the order was unpredictable. The merchants felt that they had no control, and were especially frustrated when they had made payments in advance.

The trading posts from which the British East India Company operated in India and elsewhere came to be called 'factories' because they were centres from which the Company's 'factors' or agents operated, and usually included a warehouse but rarely any manufacturing facilities. However, they were also centres through which the Company attempted to *control* the production of the textiles they were buying, and it was perhaps in this sense that the word 'factory' was eventually borrowed from the overseas context and applied to mills and workshops in Britain. This is plausible because some early factories in Britain not only housed productive processes but were centres from which hand-loom weaving or other forms of cottage industry were also organized. Meanwhile, some trading posts in India which had originally been 'factories' only in the supervisory sense gradually became more involved in manufacturing. By the 1670s, the East India Company was paying wages regularly to many weavers, and there were a few production units with up to 300 workers in each for printing cloth by hand or for reeling silk. Always, the aim was to enforce some sort of work discipline and achieve a better control of production.

It should be noted, however, that although the impetus for these developments came from European companies which wished to buy cotton and silk textiles for resale in Europe, there were Indian precedents. Thus, although the majority of India's textile output came from home-based work by people in rural areas, concentrated groups of specialist workers were to be found both in the royal factories (which continued functioning until the 1780s) and in enterprises run by Indian merchants or weavers. The latter groups were found particularly in Bengal, the centre of the finest cotton (muslin) manufacture and the silk industry. Master weavers sometimes had significant amounts of capital, owned several looms and employed other workers as 'apprentices'. In the 1750s, there were sometimes a hundred or more apprentices employed by one man, though rarely concentrated in a single building. And since at this time Indian merchants still handled two-thirds of the cloth exports from Dacca (Dhaka), the most important Bengali commercial centre, we should not assume that Europeans took all the initiatives.

Factory machines

By this time, there were not only large workshops in Britain, but there were also developments in factory machinery. To begin with, however, the situation was analogous to what we have observed in plantation agriculture. Whilst a new pattern of organization was emerging, the techniques used were often old ones, sometimes transferred from other countries.

In Britain, the pioneer mechanized factory was a mill which produced yarn for making silk stockings, set up at Derby in 1702. The mill was equipped with Dutch machines for silk 'throwing' or twisting, and they were driven by a 4-metre diameter water-wheel. This enterprise was not successful, however, until it was taken over by John and Thomas Lombe a few years later. They introduced better equipment after John had spent several months in Italy surreptitiously copying the machines used in silk mills there. On his return, Thomas Lombe took out a patent covering the use in Britain of machines built on Italian lines, and then between 1718 and 1721 set about extending the Derby mill.[15]

Another branch of silk manufactures in which the British had interests was in India. The East India Company had a 'factory' in Bengal where they purchased silk from local producers. Unwinding the silk filaments from the cocoon was a slow process which limited total output, so to speed up production they introduced reeling machines or 'filatures' from Italy in 1769. A number of Italian silk-workers were also recruited to teach the use of the filatures in India. All the machines were initially owned by the East India Company, but they were so successful that by 1800 many more were in use, owned by Indian merchants.[16]

Thomas Lombe's patent specified three types of machine (table 3), including filatures or reeling machines. The Derby silk mill was mainly processing imported raw silk from Italy, not cocoons, but even so it did have eight filatures, probably of the same type as later sent to Bengal. It will be recalled from chapter 2 that the function of reeling machines of this sort was to wind silk filaments onto a reel as they were unwound from the cocoons, in the manner shown in figure 22. This illustration comes from a book published in England in 1819, and it is probable that the reeling machine it portrays is broadly similar to the filatures used in Bengal and at Derby. There are also some striking parallels with the Chinese device shown earlier in figure 8. Both machines have a heated basin or bath of water at one end containing the silk-worm cocoons, and a large reel at the other, supported on a wooden frame. Both have a series of guides through which two filaments are drawn from cocoons in the bath onto the reel, and there is a crank mechanism to move the

Table 3 Silk-processing machines in 'transfers of technology' from Italy in the eighteenth century

Type of machine	Hand-operated or water-powered	Numbers operating		Precedents and parallels (note c)	Later adaptations for cotton spinning
		Derby: silk mill (note a)	Bengal: East India Company (note b)		
Silk-reeling machine or 'filature' (patented by Lombe, 1718)	Hand (see figure 22)	8	c.20	Chinese silk-reeling machine (see figure 8)	No comparable process with cotton
Doubler's wheel	Hand	306	See note d	Spinning wheel	—
Winding machine (patented by Lombe, 1718)	Water or hand	78	See note d	Chinese winding or spinning machine for hemp or ramie (see note e and figures 9 and 10)	Arkwright's 'water frame' patent, 1769 (see note f)
Twisting or throwing (throwsting) machine	A Square-framed type, water or hand (see figure 22)	0	See note d		
Twisting or throwing (throwsting) machine	B Circular-framed type, water-powered	4	0	Probably invented in Italy in fourteenth or fifteenth century	Paul and Wyatt circular-framed machine, 1738

a *Source*: Chaloner.

b *Source*: Bhattacharya and Chaudhuri, p. 286.

c The term 'precedent' does not imply any particular assumption about transfers of technology from earlier Chinese types.

d It is probable that the Indian type of spinning wheel, the *charkha*, was adapted for winding, doubling and twisting.

e Parallels between silk-twisting machines and Chinese hemp-spinning equipment do not extend to the bobbin-and-flyer mechanism on the former, for which the earliest parallel may be an Islamic device quoted by al-Hassan and Hill, p. 186.

f Very limited similarity (for example, in position of spindles and method of driving them).

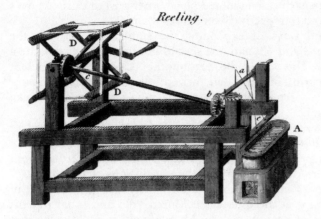

Reeling.

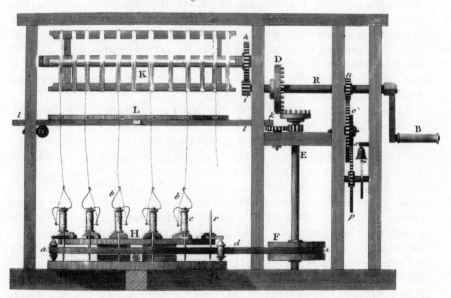

Throwsting.

Figure 22   Machines for reeling silk (top) and 'throwsting' or twisting silk
thread, as illustrated by Abraham Rees in his *Cyclopedia* of 1819.
   Both machines are shown as powered by a hand-crank, but the throwsting
or twisting machine would more usually be driven from the water-wheel in a
mill.
   (Reproduced by courtesy of Ironbridge Gorge Museum Trust.)

guides to and fro, in each case mounted on a corner post of the machine's frame (marked B and b respectively on figures 8 and 22).

There are several ways of explaining the similarities between the machines in figures 8 and 22, and one should not immediately assume a direct connection. This could be yet another case of independent invention. Also, bearing in mind that these are both nineteenth-century illustrations, they may represent a quite recent cross-fertilization between Europe and China.

Similarities between the machines are too close for independent invention to seem likely, however, and even though figure 8 may reflect nineteenth-century practice, it also fits a description dating from about 1090. It is therefore difficult to avoid the conclusions that technology was transferred from China to Italy at a fairly early date (as pointed out in chapter 3), perhaps at the time when Marco Polo and other Italians visited China in the thirteenth century. Alternatively, it could mean that silk-reeling equipment changed very little during the long process by which silk culture reached the West via the Byzantine and Islamic civilizations. It is worth noting also that there are similarities between European silk-throwing or twisting machines (figure 22) and the Chinese device for spinning hemp or ramie illustrated earlier (see figures 9 and 10). The major difference is that the throwing machine incorporates a bobbin-and-flyer device which, as noted in chapter 2, may have been invented within the Islamic or European civilizations.

As with most transfers of technology, the transfer of Italian silk machines into England stimulated local innovations, so once again we find that the ingenuity expressed in machines from one culture evoked a response elsewhere, as in a dialogue. In this instance, the Derby silk mill was of great interest to people with interests in cotton or linen spinning, and when Lombe applied for an extension of his patents in 1732 there were objections from cotton manufacturers because they 'wished to use Lombe's machines with their own particular fibres'.[17] This implies that efforts were being made to devise cotton-spinning machines modelled on the equipment at Derby. The clearest example of this was the spinning machine invented by Lewis Paul, whose idea of using rollers to 'copy' the finger motions of a human spinner has already been mentioned. Paul's machine had its bobbins and rollers mounted on a circular frame, all of them made to rotate by a large wheel turning within the frame. This unusual arrangement was very similar to the throwing or twisting machines in the Derby silk mill, which also had a circular frame (see table 3).

Cotton-spinning factories with machines of this type were set up by Lewis Paul and his partner John Wyatt in 1739, 1743 and 1748, the most important being at Northampton. The machines did not enjoy long-term success partly because, with only one pair of rollers, the quality of the yarn produced was not good. On Richard Arkwright's later and more effective machines there were three pairs of rollers, which drew out the partially spun yarn a little before the rotation of the flyer gave it a final twist. However, it appears that the ultimate failure of the Paul and Wyatt factories was more particularly a result of problems with work discipline and organization. One letter written to Lewis Paul commented that only half his employees had turned up for work that day.[18]

Richard Arkwright thus became accepted as the pioneer of cotton spinning in factories, not only because he had better machines, but also because he was more effective in tackling problems of discipline and organization. But it is also significant that Arkwright's first mills were at Nottingham and in Derbyshire, not far from the Derby silk mill. Thus the Derby mill was 'the progenitor' of the type of factory associated with Arkwright, and this is not only because of similarities in the design of the machines and the use of water-wheels to drive them, but because the Derby mill was also an example of factory organization which others had followed. Giving evidence to a Parliamentary enquiry in 1816, one Derby factory master said that local cotton-spinning mills as well as silk mills operated on the basis of a 12-hour day. Employees worked six hours before the midday meal break and six hours after. 'This has been the invariable practice at the original silk mill in Derby . . . for more than a hundred years'.[19] Thus one very important point about the machines used in this period was that they enabled the owners to impose a long working day on employees who would have worked their own hours at their own pace if using hand-powered machines at home.

Much has been made of the impact of textile machinery inventions in the eighteenth century, but it will be apparent that up to the 1760s the machines actually used in the textile industry – like the cane-crushing mills mentioned earlier on Caribbean sugar plantations – were of a very traditional kind with several centuries of history behind them. What was really new was the approach to work discipline and organization, which had parallels on plantations also, and in contemporary armies.

7 Three industrial movements, 1700–1815

Problems of resources

Between 1500 and 1750, populations almost doubled in many parts of Europe and Asia, including China, Japan, India and the west of Europe. This increase came on top of recovery from the devastating epidemics and wars of the fourteenth century (chapter 3), and took human numbers to unprecedented levels. This meant that basic requirements for food, fuel and timber also rose, and probably more than doubled. In many places, particularly in Europe and in China, the new crops introduced from the Americas contributed greatly to increased food production, especially white potatoes, sweet potatoes and maize. Indeed, it can be argued that it was the spread of these crops which allowed world population growth to accelerate. Farmers were also achieving increased yields from traditional cereals, both in the West and in China.[1]

The yield from a single crop of rice in China varied with region, variety and weather, but during the seventeenth century could range from one to three tons per hectare. During the next century, and up to 1840, cereal yields were pushed up by another 10 or 20 per cent,[2] reaching levels which were not exceeded until 1950. These increases were achieved by investing much labour in irrigation, manuring, and transplanting seedlings. At the same time, farms tended to become smaller.

Evidence of expanding output from several Chinese industries was noted in the previous chapter, and this continued in the eighteenth century. Iron was produced in large blast furnaces which were run continuously, and enterprises grew to the point where a single ironworks might employ a thousand men. Some historians[3] believe that iron output in China was greater than ever before, which means that it must have approached and perhaps exceeded 200,000 tons per year. After 1700, there was a growth of small workshops producing telescopes and mechanical novelties copied from western imports. Textile and pottery production were also expanding.

Despite the innovation and enterprise which this implies, by 1800 the Chinese economy was in difficulty. One provincial governor who was particularly knowledgeable on copper and silver working in Yunnan wrote to the imperial government in 1844 commenting on bureaucratic interference with the mines. However, it is likely that the biggest problem was that, as population growth continued and food production necessarily increased, more land was used for food crops and less was growing cotton, timber, or fodder for animals. Raw cotton was increasingly imported from India, but shortages of timber were not adequately offset, even though metals were being extensively worked, often with coal-based fuels. There were practical difficulties in the way of totally eliminating dependence on charcoal fuel. However, it is perhaps characteristic of a society with such sophisticated agriculture, and a leaning toward horticultural interests among educated people, that less progress was made with industrial methods of overcoming shortages than with the expansion of food production. Allowing for payments in kind and trade by barter, one can estimate that the result of these conditions was that food remained relatively cheap compared to other costs, which meant that the effective wage paid to a labourer could also remain low. By contrast, it became comparatively more expensive to buy a cart, or build a machine, or keep animals for pulling ploughs and carrying loads. Thus it may be that after 1800, there was a tendency to use more human labour for cultivating the land, for pumping water, for textile production and even for transport of goods. Water-powered machines such as those made for the thirteenth-century textile industry (see figures 9 and 10) were now less often built.

One factor which made the situation in Europe very different was that food prices and wages were never so low relative to other materials, and there was a much greater incentive to use machines. More fundamentally, however, innovations in the use of coal together with the introduction of the steam engine allowed new resources to be exploited, and this constitutes the first of three industrial movements of the period through which the West gained a decisive advantage with respect to material production. The limits to development resulting from reliance on land-based (renewable) resources were slowly removed. Before that transition was complete, it was often a scarcity of firewood or timber which was the most critical resource shortage. Charcoal was the essential fuel of many industries, including iron-smelting, and the cutting of trees and the charring of timber in closed heaps (figure 23) was an important activity in many industrial areas.

It was in Britain that deforestation had gone furthest, but the picture is a complex one. Although shortage of large timber suitable for shipbuilding undoubtedly became acute during the eighteenth century, charcoal fuel for the iron industry was produced by cutting young growth from 'coppiced' trees. This would regrow in less than ten years, so by using woodlands in rotation, production could be sustained indefinitely.

Figure 23 Making charcoal by charring wood in closed heaps.
These are shown at various stages from a newly ignited heap (left foreground) to collapsed heaps in which the process is nearly complete (behind). A man is cutting saplings for use in a new batch (left background).
(From the *Universal Magazine*, 1747, by courtesy of Ironbridge Gorge Museum Trust.)

Thus it was not so much a shortage of charcoal for *existing* furnaces which was a problem, but the difficulty of finding extra woodland to support the new furnaces required by an expanding industry. The way round this difficulty was found by increasingly using coal instead of firewood, or coke instead of charcoal. The shortage of timber for ships was dealt with partly by importing wood, partly by building ships abroad, and ultimately, from the 1830s, by more often building ships of iron.

Steam engines, iron and coal[4]

It is possible to argue that Britain and China – and several other countries also – were facing comparable (but not identical) problems of rising population and shortage of resources in the eighteenth century. Why one country was more successful in solving those problems than another is not easy to understand. Britain's achievement in this respect did not initially arise from more 'advanced' technology, but from an unusually open attitude to ideas from elsewhere, and a vigorous technological dialogue. With regard to textile industries, this entailed dialogue with Asia as well as with Italian silk producers (chapter 6). With regard to the invention of the steam engine, key ideas came from Italy and Germany (especially discoveries about atmospheric pressure), and from the Frenchman Denis Papin, who invented a steam engine of sorts in the 1680s (but did not perfect it as an economically useful machine). A discussion of how these ideas led to further inventions in England, culminating in Newcomen's steam engine of 1712, has been provided in another book.[5] One point to note here is that not only did early steam engines in England burn coal under their boilers, not wood, but it was at coal mines that nearly all of them were used, mainly for pumping water (figure 24).

For over a century, the price of firewood in England had consistently increased at twice the rate of most other prices, and coal had been more and more widely used as a fuel. Domestic grates and chimneys had been adapted for coal burning, and an effort was being made to adapt industrial processes. In some instances, coal was only suitable as a fuel if it was first converted into coke, and in 1709, Abraham Darby found that coke could be used instead of charcoal to fuel blast furnaces.

Once the technical difficulties associated with this innovation had been overcome, it was clear that the coke-fired blast furnace would release the British iron industry from what could have become a very confining limitation, namely dependence on charcoal. The introduction of coke-fired iron-smelting in England is therefore regarded as an important milestone, and the place of its introduction, at Coalbrookdale in Shropshire, has been hailed as 'the birthplace of the industrial revolution'. It was some time before the success of the coke-blast furnace was as great as this terminology suggests, however, because quality control proved difficult. Only after about 1760 did cast iron made with coke become a real success, and then the expansion of this branch of the industry was very rapid.

In China, several centuries earlier, the use of coke-fired blast furnaces had also enabled a rapid expansion of iron production to take place in the Hebei area, as described in chapter 1. It has to be stressed now that the

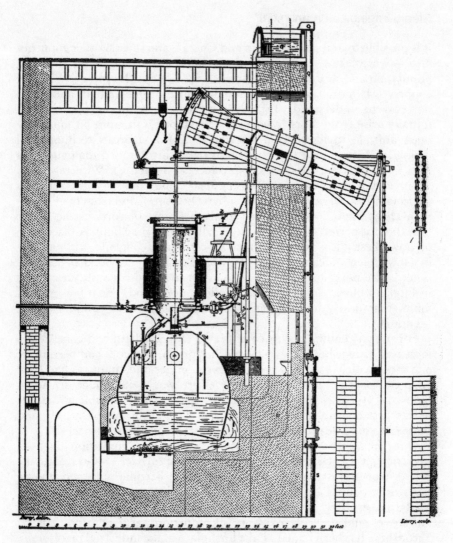

Figure 24 A steam engine of the type invented by Thomas Newcomen in 1712, as improved by John Smeaton in the 1770s.

This engine was designed to pump water from a coal mine in northeastern England. The boiler and cylinder are to the left of centre. The top of the mine shaft is on the right, with a heavy rod down the middle of it to connect the engine to pumps in the mine below.

(From John Farey, *A Treatise on the Steam Engine*, London, 1827; reproduced by courtesy of Ironbridge Gorge Museum Trust.)

coke-blast furnace developed by Abraham Darby in 1709 was an entirely *independent invention*. Darby could have had no knowledge of the Chinese process. The common factor was simply that in both locations, supplies of charcoal from woodlands set limits on iron output. As so often with independent inventions, then, a specific environmental pressure evoked similar responses in quite different places, and at different dates.

One obvious reason why the Chinese coke-blast furnace had a less 'revolutionary' impact in China than in Britain was that it was not complemented by the steam engine. By 1734, there were up to a hundred of these machines working in the coal mining regions of England, pumping water from mines. Some coal seams which had been out of reach because of flooding could now be worked. It was certainly fortunate for the development of the engine that it was so badly needed at mines, because at first it was only at a coal mine that a steam engine could be economically operated. The thermal efficiency of the Newcomen engine was only about 0.5 per cent, and it used a prodigious amount of fuel. Another limitation was that the engine was designed solely for use as a pumping device, so the tiny handful of engines not used in mines were employed in waterworks of one sort or another, notably in London and at a palace in Vienna.

Despite these restrictions, close links with the iron industry were quick to develop. One of the most demanding aspects of building a steam engine was making the cylinder. It had to be cast in metal, and presented some of the same difficulties as casting bells or large cannon. These were at first made of bronze, and only later was it possible to cast them in iron. In the same way, steam-engine cylinders were at first made of brass, but in 1718 a cast-iron cylinder was made at Coalbrookdale. More were made in the 1720s, and by 1731 the casting of iron engine cylinders was a regular part of the Coalbrookdale business.

In this region, too, the pumping action of the steam engine was first used to provide the air blast to furnaces. It will be recalled that Chinese blast furnaces depended on manually powered bellows (see figure 1), with only occasional use of water-wheels. The norm in Britain was for furnaces always to be blown by water-powered bellows, and as a result operations often had to stop in dry weather. At Coalbrookdale in 1742, a steam engine was employed to pump water so that water-powered bellows could be kept going at all times. Then in the next few years air pumps with iron cylinders were introduced to replace bellows, and from 1776, again in Shropshire, there were machines of this sort driven directly by steam engines. This liberated the blast furnace from dependence on water power, and removed another constraint which would otherwise have limited the production of iron. Thus steam engines developed in symbiosis with the iron industry as well as with coal mining. Both industries needed the engine for their expansion, and at the same time contributed to its progress.

During the middle decades of the eighteenth century, while the imperfections of the early coke-blast process were being sorted out, and before James Watt had begun his radical work on the steam engine, Britain's timber and fuel shortages had to be dealt with by imports from countries with more plentiful timber and charcoal resources. Roughly half the iron used in Britain came from Sweden and Russia. As noted in chapter 5, Russia had been building up a considerable iron industry for some time, and by about 1780 it was the world's leading exporter of iron.

Meanwhile, the shortage of timber for shipbuilding was dealt with partly by imports of timber, again from Sweden and Russia, and also from America, and some shipowners had vessels built abroad. In 1774, it was said that one-third of all new British shipping was built on America's eastern seaboard, though this practice was soon interrupted by the American War of Independence. Meanwhile, the British East India Company was regularly building many of its vessels in India, and during the Napoleonic wars the Royal Navy began to have warships built there also.

In varied ways, therefore, both by technical innovation and through the involvement of other countries, the limits to economic growth which would have been imposed on Britain by reliance on land-based resources were overcome. However, it has been argued that resource shortages were still felt in Britain until about 1820. Only after that date could living standards rise decisively for the majority of Britain's increasing population.

Estimates of iron production per capita (that is, per person in the total population) have been used to show the breakthrough to higher levels of production which became possible when the iron industry freed itself of limitations imposed by dependence on charcoal fuel,[6] and we may add, dependence on water power (table 4). Whilst most pre-industrial economies produced less than 2 kilograms of iron per capita in a year, western Europe, with Britain ahead of other countries, had far exceeded that figure by 1800. The statistics from Russia demonstrate what could be achieved by a very traditional industry, where charcoal and water-power resources were plentiful. However, it must be remembered that much of this iron was exported. Britain and France were producing iron on a similar scale for their own use.

Table 4 Estimates of iron production per head of population

Standard units are metric tons of pig-iron equivalent, but the following figures are 'broad-brush' estimates in which discrepancies in units or pig-iron conversion factors make very little difference.

Country or region	Date	Total iron production (tons per year)	Iron production per head of population (kg per year)
Examples of early production levels			
China (Song Empire: see chapter 1)	1078	125,000[a]	1.4[a] (or more)
Europe (excluding Russia)	1500	60,000[b]	1.0[b] (maximum)
Eighteenth-century production levels			
China	c.1750	200,000[c]	1.0[c] (±0.4)
India	c.1750	200,000[d]	1.0[d] (±0.5)
Russia	1793	202,000[e]	5.0[e] (much lower in 1750)
Europe (excluding Russia)	1700	175,000[a]	1.8[a]
	1750	200,000[b]	1.5[f]–1.7[f]
	1796	420,000[a,g]	4.0
	1806	700,000[h]	6.0[h]

Sources: [a] Hartwell; figures for China may be underestimates due to omissions from official statistics.

[b] Boserup, p. 107, quoting the higher and more realistic end of her range.

[c] See text and comments by Elvin, also Wagner.

[d] Dharampal, p. LIV.

[e] Esper.

[f] Boserup's higher iron production figure, recalculated using her own population estimates for 1750.

[g] B. R. Mitchell, pp. 747, 773, interpolated where necessary, including British production of 127,000 tons.

[h] B. R. Mitchell, including British production (now 248,000 tons) and French production (around 200,000 tons).

A second industrial movement

The effect of this change in the resource base on which iron production depended was so dramatic that it is certainly justifiable to celebrate the small area in Shropshire where so many key innovations originated. But the industrial revolution was more than just a shift from wood-based fuels to coal, from timber construction to iron, and from water-wheels to steam power. There was also a second industrial movement of great importance, related to the ideas about *organization* of production mentioned in the previous chapter. These ideas certainly had their influence in Shropshire, where 'new forms of work discipline' were enforced by iron-masters and magistrates, and were encouraged by evangelical religion.[7]

However, the classic expression in Britain of the new discipline was the textile industry, represented here by cotton spinning, dyeing and calico printing. In the spinning mills, especially, strict hours and machine-paced work complemented one another. The origins of the textile factory were discussed in chapter 6, where it was noted that the first British examples included the silk mill at Derby and Richard Arkwright's earliest mills nearby. Thus if Shropshire was the cradle of one sort of industrial movement, addressed to problems of resources, Derbyshire was the cradle of another, much more concerned with factory organization. It is worth representing these as two quite separate industrial 'movements' because it was some time before they coalesced to create 'the industrial revolution'.

The difference between these two movements is particularly clear when we notice how traditional the early factories were in the resources on which they depended both for power (water-wheel and sometimes even horse-gins) and for construction of machines (timber, with a few iron parts made by blacksmiths or, where gear-wheels were needed, clockmakers). Indeed, so traditional was the construction of these factories that we should not be surprised that some of the machinery they contained was little different from earlier Islamic or Chinese equipment (chapter 6). There was nothing in early silk and cotton mills that would have surprised or puzzled a millwright from the medieval Islamic civilization, or from China, except their enormous size and their organization.

However, in 1782 and 1784 James Watt took out patents for the crucial inventions which enabled steam engines to be satisfactorily used for the first time to power factory machinery, and in the same decade cast-iron columns were introduced into many cotton mills to support floors sagging under increasingly heavy machines. In 1796, iron beams were first used in a factory building, and by 1800 iron was increasingly used for wheels, shafts and frames in machines. Water-wheels were still an important source of power, but iron construction allowed larger and more efficient wheels to be made.

So whilst in its early years the British textile factory had chiefly represented a revolution in production methods, after 1785 it increasingly reflected the revolution in sources of power and materials also. Moreover, the growing predominance of cotton in the British textile industry reflected another sort of revolution. Linen and wool were produced in the British Isles and there had long been a textile industry based on these fibres. But cotton could not be produced in a north European climate, so the growing cotton industry was entirely dependent on an imported raw material, and on unfamiliar adaptations of basic textile technologies for spinning yarn, dyeing cloth and other processes. Although some raw cotton came from India and elsewhere, the development of cotton as a plantation crop in America paralleled the growth of the British industry very closely.

However, the question of how the British acquired the technology to deal with these relatively unfamiliar fibres is a complex story. With silk, as we have seen, the technology could be copied ready-made from Italy. Then silk-twisting machines gave some impetus to the development of equipment for cotton spinning (chapter 6).

Another element in the story is undoubtedly the ingenuity of a few mechanically minded people in the textile areas, their observation of how cotton behaved when spun with a traditional wheel, and their awareness of the buoyant demand for cotton yarn, which was being used by weavers almost faster than it could be spun. One invention which may have arisen from circumstances of this sort was the Hargreaves spinning jenny of the 1760s.

Asian stimulus and the second industrial movement

A further aspect of industrial development in textile production was the challenge presented to British manufacturers by imported Indian cloth, both in terms of the fineness of muslins from Bengal, and with respect to the brilliant colours and patterns of other cloths. It was especially striking that the colours did not fade when the cloth was washed, as happened all too often with European fabrics. Cotton yarn spun in Britain, even on Arkwright's machines, was never fine nor strong enough for fine muslins to be woven. But British spinners showed little interest in enquiring how their Indian counterparts achieved this high quality, and would have been disappointed if they had, for it was done mainly by very painstaking and laborious hand-spinning. Thus there was no transfer of spinning technology, yet there was still a dialogue in the sense that the Indian product demonstrated what could be achieved, and provided a continuing stimulus to inventors of machinery. With Samuel Crompton's invention of the spinning 'mule' at the end of the eighteenth century, that stimulus had its fulfilment, and it was at last possible to equal and surpass the quality of the Indian product.

Some of the most colourful and attractively patterned Indian cloth to be sold in Europe before 1700 – and before England and France placed restrictions on its importation – came from the eastern coast of the subcontinent, notably from around Pondicherry. Here, cloth was usually hand-painted with dye or mordant, and the dyeing processes were often very intricate. When efforts were made to reproduce these processes in Europe, there was immediate need for a quicker method of production. Thus, whilst there was a strong stimulus from the high quality cloths of the Pondicherry region, the techniques actually copied in Europe were based on printing the cloth rather than painting it, using 'Persian' wood blocks, or the closely related block-printing methods of Gujarat (figure 25). Historians of textile technology note that Europeans 'retained many features of the Oriental technique', including the mordants used with madder, and wax-resist printing with indigo.[8]

It must be stressed that wood-block printing on cloth had been

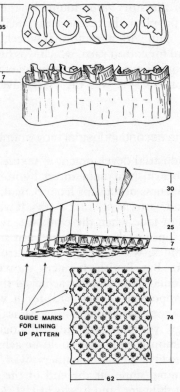

Figure 25   Wood blocks for printing cloth in India, with a perspective view of each, and also a view of the printing face.

Dimensions are in millimetres. The top illustration shows a block carved with letters in the Persian script, laterally inverted.

(Drawn from examples collected by the East India Company after 1800, now in the Victoria and Albert Museum, London.)

practised on a limited scale in Europe for several centuries. What was new was adaptation of the technique for the bright colours and fast dyes of India and Persia. The first successful attempts to imitate the Indian imports were made almost simultaneously in Britain, Holland, France and Switzerland in the 1670s. One important early printworks was that of William Sherwin, on the River Lea in London.

By the 1740s, so much experience had accumulated that the owner of a printworks in Basle could respond to an account of Indian dye techniques by saying that most of them were already used in Basle, as if there was nothing more to be learned from Asia. In the next decade, the French government relaxed an earlier prohibition on printed cloth, and several new factories developed. One of the biggest was at Orange, where over 17,400 lengths of cloth were printed in 1762; others were at Troyes and Rouen. In all these works, processes akin to Indian techniques were used, often modified to save labour, but there was still dissatisfaction that the brilliant shades of red seen on Indian cloth could not be reproduced.

It was recognized that Turkish dyeing methods, which were closely related to the Indian ones, could achieve the shades of red required. Thus a dyer from Manchester named John Wilson sent a young man to Ismir (Smyrna) in Turkey to 'procure . . . the secret of dyeing Turkey red'. Wilson spent a considerable amount on this project, since the young man stayed long enough in Smyrna to learn the language. Finally, in 1753, he 'got Admittance into their Dye-house, and was instructed', but the process seemed too slow and labour-intensive to be economical in England.[9]

Wilson (and other British dyers) next turned for advice to the French producers of high-quality dyed cloth, but found that their best information had come from Turkey also, as a result of Greek and Turkish dyers emigrating into France during the 1740s, some of them settling at Rouen. Meanwhile, new techniques for printing cloth were being developed, including printing from copper plates in a press, which allowed more intricate patterns to be reproduced than with wood blocks. Some of the best work of this kind was done in London and at Jouy, near Paris, with some of the equipment used in France being imported from England in the 1770s. Another innovation was printing from rollers or cylinders, which allowed the whole process to be speeded up. This had roots in both Switzerland and Britain. It became the basis for the mass printing of cheap textiles on the continent (at Mulhouse), and in England was used in the Lancashire textile area from 1785.

These frequent exchanges between British, French and Swiss dyers and printers are indicative of a vigorous technological dialogue. Despite the innovation which resulted, however, the secret of printing bright reds still eluded British dyers. Their answer came when several Frenchmen, accompanied by a dyer of Turkish origin, moved to the Manchester area, where one set up a Turkey-red dyeworks in 1781. Meanwhile, a Scottish associate of Richard Arkwright persuaded a French dyer named Papillon to start a Turkey-red enterprise near Glasgow in 1785.

The timing of these developments is related to a rapid increase in the production of 'calicoes' in Britain. This term comes from the name of the South Indian port of Calicut, and refers to a less fine cloth than muslin. For the manufacture of calicoes, yarn spun on Arkwright's machines proved suitable. Arkwright had been lobbying the government, and had persuaded it to reduce the excise duty payable on calicoes made in Britain. In 1775, the first year after this reduction in duty, 1,930 thousand metres of calico were imported, mostly from India, whilst only 52 thousand metres were made in Britain. By 1783, imports had fallen by two-thirds, whilst British production had risen dramatically. There was thus a massive increase in the amount of cotton cloth needing to be dyed, and it was natural that dye and print factories were set up close to the areas where calicoes were being produced in Lancashire and Scotland.

One other source of innovation in this industry was scientific study.[10] Several French chemists had attempted crude scientific theories of dyeing, and one minor discovery was the role of calcium minerals in the madder process, as mentioned in chapter 6. The greatest of these chemists was Berthollet, who developed the chlorine bleaching process during the 1780s. Historians who are sceptical of the role of science in the industrial revolution point out that chemistry was in a rudimentary state, and suggest that chlorine bleaching was its only real contribution. But the habit of writing down experimental results, and the tendency to conceptualize processes by proposing theories, however inadequate, suggested new approaches and led to a build-up of more and better-arranged empirical knowledge. Thus processes which had been used for centuries in India, Iran and Turkey were extended quite rapidly with many new applications.

A third industrial movement

Although the Indian cotton trade looms vaguely in the background of many histories of the industrial revolution, Fernand Braudel is one of very few writers on the period who had recognized that India was a creative influence of some significance for British technology.[11] One aspect of this influence was the low cost and high quality of Indian cloth. Labour was plentiful in the Indian textile areas and wages were low. There was little incentive, therefore, for Indian merchants to mechanize production. As Braudel puts it, the incentive 'worked the other way round'. New machines were invented in Britain to try and equal Indian cloth in both cheapness and quality, and there were the transfers of dyeing techniques just discussed.

Nevertheless, technological changes occurred in India associated with the impact of British trade. Moreover, the linkages were sufficiently strong for us to speak of a third industrial movement, located in India[12] but related to the two British movements already described. It comprised expansion in the textile, chemical and shipbuilding industries. In textiles, despite lack of any general mechanization, silk-reeling machines were introduced in Bengal (chapter 6). Some master weavers in the same region invested in extra looms, and a few large weaving establishments developed.

Textile production depended on a range of chemicals, including vegetable dyes and mineral substances used as mordants. In terms of the volume of output, production could be on a considerable scale, but again with very little capital investment. For example, alum was needed as a mordant in madder dyeing, and was produced in Rajasthan by processing broken shales tipped as waste around copper mines. The first stage was to steep the shale in water in a series of earthenware pots. Vast numbers of these pots were used, and their cost and the cost of building a 'boiling house' must have been the only capital involved in the process. The pots were arranged in long rows on ledges dug into the sides of the shale tips (figure 26). Each pot of shale was subjected to three changes of water, the water being poured from one pot to another until it was a dirty blue colour. It was then taken to the boiling house. Some water was evaporated off, and the pot was left to stand. Blue crystals of copper sulphate formed and were removed. After decanting the liquid and boiling it again, saltpetre was added. The alum was formed by reaction of the saltpetre with aluminium salts in the solution, and crystallized at the bottom of the vessel.

An industry operating this process with about 50 boiling houses developed at Khetri and Singhana in Rajasthan. A notable feature was the simplicity of the equipment used. This was paralleled in textile painting around Pondicherry and calico printing at Ahmedabad. In the first instance, cloth being painted was spread on the ground in the open, whilst at Ahmedabad and elsewhere the washing of cloth between dyeing and printing processes was done in the river. Minimal equipment also contributed to low costs, and again meant that there was little extra profit to be had by investing in production. By contrast, investment in trade could be very lucrative, because cloth made cheaply in India was sold for three or four times the price in overseas markets, and there were plenty of people in eighteenth-century India with funds to invest. They

Figure 26 Alum production in Rajasthan, India, about 1800.

Shale tips associated with copper mines provided the raw material and can be seen in the background on the right. Shale was steeped in water in rows of pots on the tips. The liquid was then brought to the boiling house seen to the left.

(Illustration by Hazel Cotterell, developed from information given by Ray.)

included the great banking firm of Jagar Seth, and the many Parsi and Gujarati bankers to be found in ports such as Surat and Bombay. European companies sometimes took loans from these firms, and in the financial crisis of 1720–1 associated with the South Sea Bubble, the British East India Company saw that it could safeguard a shaky position in London by borrowing money 'in India at interest'.

The one branch of technology which benefited from the tendency to invest in trade rather than production was shipbuilding. Thus the 'industrial movement' in India of most importance during this period relates particularly to the main shipbuilding centres (figure 27). Ships were built for three types of customer: merchants, local rulers and European companies. Whilst small vessels for the coastal trade were of purely local design, most of the larger ships built in the eighteenth century were superficially similar in hull and sail-plan to European vessels. However, there were characteristic differences in the design of the bows and other detail, and there was also a distinctive carpentry technique for making rabet-work seams between planks.

This merging of Indian carpentry techniques with European design reflects a cosmopolitan approach which had been very characteristic of technology in India for a long time. For example, assessments of Mughal development in the sixteenth century tend to present Emperor Akbar's policy for encouraging trade and industry as forward-looking, but regard him as short-sighted in not seeing the need of formal arrangements for training craftworkers and engineers.[13] Indian rulers neglected this because they found it too easy to employ skilled people from other countries, often as mercenaries in their armies. There were also many Armenians and Persians in the textile trade and Arabs in shipping. Similarly, when Indian merchants invested in ships (which always carried guns for defence against pirates), they often recruited Europeans to captain the ships or as gunnery officers.

When it came to building ships, however, India had a great deal of local expertise, and few Europeans were involved. This is clear from the history of Bombay dockyard, where instead of Indian merchants being dependent on European skills, we can observe a European company depending on Indian skills. The dockyard belonged to the British East India Company, which had established Bombay as its base in western India in 1686. At first, only small vessels were built there, but in 1736, when an expansion of the shipbuilding programme was proposed, the Company found themselves short of carpenters, and recruited skilled men from Surat. For four years they worked under the direction of Robert Baldry, but then he died and one of the Surat carpenters, Lowjee Wadia, was appointed in his place and took on a very considerable

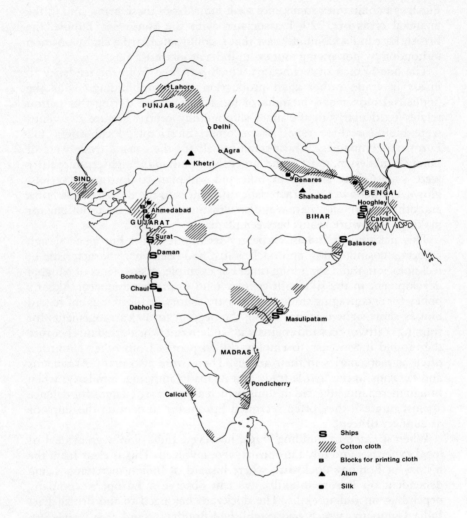

Figure 27 Shipbuilding centres and the location of handicraft industries connected with the export of textiles from eighteenth-century India.
(Based on information given by Chaudhuri, also Irwin and Brett, Mookerji, Ray.)

responsibility. For example, when a dry dock was constructed in 1750 for repairing ships, it was Lowjee who chose its site. The dock was 64 metres long, and was twice extended under Lowjee's supervision.

When he died in 1774, Lowjee was succeeded as Master Carpenter by his eldest son, with another son as assistant. Thereafter, all Master Carpenters at Bombay were members of the Wadia family until 1885. Initially the dockyard was important mainly for repair work, and only small vessels were built there. However, in 1768, a ship of 500 tons was begun, and in 1778, a vessel of 749 tons was launched. When it reached London the following year, the Company was so pleased that they ordered another of the same design.

By the 1790s, timber shortages in Britain were encouraging the construction of more and more large ships at Bombay. In addition, with the Napoleonic wars in progress, the British Royal Navy needed extra ships, and estimates were sought from Bombay as to the cost of building warships there. Rear-Admiral Sir Thomas Trowbridge wrote directly to Jamsetji Wadia in 1802 commenting that shipyards in Bengal had also sent estimates, but he had 'no opinion' of the craftworkers there. Jamsetji's reply to the Rear-Admiral evidently inspired confidence, as did the Royal Navy's experience when they purchased a Bombay-built ship in 1804. Thus the original idea of sending out an English shipwright to supervise construction was given up as unnecessary. The Royal Navy simply sent out drawings for Jamsetji's team to work from. During the next twenty and more years a total of 22 ships was built for the Royal Navy at Bombay. In service, they achieved a great reputation for durability. In the winter of 1809–10, one of them, HMS *Salsette*, was the only ship to survive crushing by ice when an expeditionary force in the Baltic Sea got 'frozen in'. Equally impressive is the record of HMS *Trincomalee* (later renamed *Foudroyant*), which is still afloat after a life of 170 years.

Statistics on ships built at Bombay between 1736 and 1859 are presented in table 5. Since the shipyard was owned by the East India Company, it is understandable that a large proportion of the vessels built there was for them. Many boats were also built for the Bengal Pilot Service, which served the Company's interests. Spare capacity was then used to build for the Royal Navy, for the navy of the Sultan of Muscat (in Arabia) and for private owners, many of whom were local merchants. For example, three of the ships built in the 1780s were for Parsi merchants in Bombay.

There were several technical developments during the fifty years up to 1815, one being the disappearance of Surat rabet work, possibly as a result of the introduction of new methods (and tools) for sawing planks, but possibly also because the Royal Navy insisted on its own standards.

Table 5   Numbers of ships built on the Hooghley and at Bombay in successive 20-year periods
The figures exclude vessels of less than 100 tons displacement. After 1790, many ships were over 1,000 tons, among them 15 built at Bombay between 1801 and 1821.
(Compiled from data given in different form by Mookerji and Wadia.)

Period	Numbers of ships	
	Hooghley shipyards (Mookerji's data)	Bombay shipyard (Wadia's data)
1736–1760	0[a]	26
1761–1780	n.d.	40
1781–1800	35	35
1801–1821[b]	237	58
1822–1840[b]	104	43[c] + 5[d]
1841–1860	0[a]	11[e] + 13[d]

n.d. = no data
[a] No large ships built, but details lacking.
[b] Departure from precise 20-year periods; dates are inclusive.
[c] Includes five steamships with conventional Indian-built timber hulls but engines from Britain, the first being HCS *Hugh Lindsey*, 411 tons, launched 1829.
[d] Iron steamers, mostly river-boats, assembled from parts sent out from Britain, the first in 1838.
[e] Includes nine steamships with timber hulls, mostly over 1,000 tons displacement. Large numbers of coal barges and other small vessels also constructed.

Increasing amounts of ironwork were being used in ships, most of which was produced in India. Ships were usually built on slipways, as in Britain, but some work on new ships was also done in the excellent dry dock at Bombay (figure 28).

The only other Indian shipyards for which statistics are available are those on the Hooghley River in Bengal. Once again the East India Company was the dominant influence, but there were also many privately owned ships plying from nearby Calcutta to Asian ports and to London.

Figure 28   HMS *Trincomalee* in the dry dock at Bombay in 1817.
Although most ships were built on slipways, the *Trincomalee* is known to have been in the dry dock just before the masts were erected. It is shown here with the temporary flagpoles put up for its launch or 'floating out' from the dock.
(Illustration by John Nellist based on photographs of the dock and studies of the ship as it currently exists, made with the help of Hartlepool Ship Restoration Company.)

Some of the largest ships built in the region and intended for long voyages conformed to the European schooner rig.

The figures in table 5 make it clear that both in Bombay and on the Hooghley output peaked in the first two decades of the nineteenth century and then declined. This reflects the influence of the Napoleonic wars in Europe, where the price of timber for shipbuilding reached a maximum between 1806 and 1813. During these years, then, ships built in India had a maximum cost advantage whilst after the wars ended in 1815, the Royal Navy needed fewer ships from Bombay (and ordered no more after 1829). However, the sharp decline in Indian shipbuilding after 1820 reflects a changed political situation also.

Deindustrialization[14]

During the eighteenth century, India participated in the European industrial revolution through the influence of its textile trade, and through the investments in shipping made by Indian bankers and merchants. Developments in textiles and shipbuilding constituted a significant industrial movement, but it would be wrong to suggest that India was on the verge of its own industrial revolution. There was no steam engine in India, no coal mines and few machines. The map (see figure 27) also shows that the expanding industries were mostly in coastal areas. Much of the interior was in economic decline, with irrigation works damaged and neglected as a result of the breakup of the Mughal Empire and the disruption of war. Though political weakness in the empire had been evident since 1707, and a Persian army heavily defeated Mughal forces at Delhi in 1739, it was the British who most fully took advantage of the collapse of the empire. Between 1757 and 1803, they took control of most of India except the Northwest. The result was that the East India Company now administered major sectors of the economy, and quickly reduced the role of the big Indian bankers by changes in taxes and methods of collecting them.

Meanwhile, India's markets in Europe were being eroded by competition from machine-spun yarns and printed calicoes made in Lancashire, and high customs duties were directed against Indian imports into Britain. Restrictions were also placed on the use of Indian-built ships for voyages to England. From 1812, there were extra duties on any imports they delivered, and that must be one factor in the decline in shipbuilding documented in table 5. A few Indian ships continued to make the voyage to Britain, however, and there was one in Liverpool Docks in 1839 when Herman Melville arrived from America. It was the *Irrawaddy* from Bombay and Melville commented: 'Forty years ago, these merchantmen were nearly the largest in the world; and they still exceed the generality.'

They were 'wholly built by the native shipwrights of India, who ... surpassed the European artisans'. Melville further commented on a point which an Indian historian confirms,[15] that the coconut fibre rope used for rigging on most Indian ships was too elastic and needed constant attention. Thus the rigging on the *Irrawaddy* was being changed for hemp rope while it was in Liverpool. Sisal rope was an alternative in India, used with advantage on some ships based at Calcutta.

Attitudes to India changed markedly after the subcontinent had fallen into British hands. Before this, travellers found much to admire in technologies ranging from agriculture to metallurgy. After 1803, however, the arrogance of conquest was reinforced by the rapid development of British industry. This meant that Indian techniques which a few years earlier seemed remarkable could now be equalled at much lower cost by British factories. India was then made to appear rather primitive, and the idea grew that its proper role was to provide raw materials for western industry, including raw cotton and indigo dye, and to function as a market for British goods. This policy was reflected in 1813 by a relaxation of the East India Company's monopoly of trade so that other British companies could now bring in manufactured goods freely for sale in India. Thus the textile industry, iron production and shipbuilding were all eroded by cheap imports from Britain, and by handicaps placed on Indian merchants.

By 1830, the situation had become so bad that even some of the British in India began to protest. One exclaimed, 'We have destroyed the manufactures of India', pleading that there should be some protection for silk weaving, 'the last of the expiring manufactures of India'. Another observer was alarmed by a 'commercial revolution' which produced 'so much present suffering to numerous classes in India'.[16]

The question that remains is the speculative one of what might have happened if a strong Mughal government had survived. Fernand Braudel argues that although there was no lack of 'capitalism' in India, the economy was not moving in the direction of home-grown industrialization. The historian of technology inevitably notes the lack of development of machines, even though there had been some increase in the use of water-wheels during the eighteenth century both in the iron industry and at gunpowder mills. However, it is impossible not to be struck by the achievements of the shipbuilding industry, which produced skilled carpenters and a model of large-scale organization. It also trained up draughtsmen and people with mechanical interests. It is striking that one of the Wadia shipbuilders installed gas lighting in his home in 1834 and built a small foundry in which he made parts for steam engines. Given an independent and more prosperous India, it is difficult not to believe that a response to British industrialization might well have taken the form of a spread of skill and innovation from the shipyards into other industries.

As it was, such developments were delayed until the 1850s and later, when the first mechanized cotton mill opened. It is significant that some of the entrepreneurs who backed the development of this industry were from the same Parsi families as had built ships in Bombay and invested in overseas trade in the eighteenth century.

8 Guns and rails: Asia, Britain and America

Asian stimulus

Britain's 'conquest' of India cannot be attributed to superior armaments. Indian armies were also well equipped (figure 29). More significant was the prior breakdown of Mughal government and the collaboration of many Indians. Some victories were also the result of good discipline and bold strategy, especially when Arthur Wellesley, the future Duke of Wellington, was in command. Wellesley's contribution also illustrates the distinctive western approach to the organizational aspect of technology. Indian armies might have had good armament, but because their guns were made in a great variety of different sizes, precise weapons drill was impossible and the supply of shot to the battlefield was unnecessarily complicated. By contrast, Wellesley's forces standardized on just three sizes of field gun, and the commander himself paid close attention to the design of gun carriages and to the bullocks which hauled them, so that his artillery could move as fast as his infantry, and without delays due to wheel breakages.

Significantly, the one major criticism regularly made of Indian artillery concerned the poor design of gun carriages.[1] Many, particularly before 1760, were little better than four-wheeled trolleys (see figure 29). But the guns themselves were often of excellent design and workmanship. Whilst some were imported and others were made with the assistance of foreign craftworkers, there was many a brass cannon and mortar of Indian design, as well as heavy muskets for camel-mounted troops. Captured field guns were often taken over for use by the British, and after capturing 90 guns in one crucial battle, Wellesley wrote that 70 were 'the finest brass ordnance I have ever seen'.[2] They were probably made in northern India, perhaps at the great Mughal arsenal at Agra.

Whilst Indians had been making guns from brass since the sixteenth century, Europeans could at first only produce this alloy in relatively small quantities because, as explained in chapter 5, they had no technique for smelting zinc. By the eighteenth century, however, brass was being produced in large quantities in Europe, and brass cannon were being cast at Woolwich Arsenal near London. Several European countries were

131

Figure 29   Indian weapons in the second half of the eighteenth century.
These include (from foreground to background): a disused cannon barrel, a
long matchlock musket fired from a sitting position, a heavy musket for use
by camel-mounted troops, brass cannons, and to the right of them a mortar.
The gun carriages shown are not typical, but similar ones were used in the
1760s, and the mortar is also of that date.
(Illustration by Hazel Cotterell. Sources include Egerton on numbers and types of
weapons in major battles, and pictorial material in the book by Pant and at the 'Armies
of India' exhibition, National Army Museum, London, 1987.)

importing metallic zinc from China for this purpose. However, from 1743 there was a smelter near Bristol in England producing zinc, using coke as fuel, and zinc smelters were also developed in Germany. At the end of the century, Britain's imports of zinc from the Far East were only about 40 tons per year. Nevertheless, a British party which visited China in 1797 took particular note of zinc smelting methods.[3] These were similar to the process used in India, which involved vaporizing the metal and then condensing it. There is a suspicion that the Bristol smelting works of 1743 was based on Indian practice, although the possibility of independent invention cannot be excluded.

A much clearer example of the transfer of technology from India occurred when British armies on the subcontinent encountered rockets, a type of weapon of which they had no previous experience (figure 30). The basic technology had come from the Ottoman Turks or from Syria before 1500, although the Chinese had invented rockets even earlier. In the 1790s, some Indian armies included very large infantry units equipped with rockets. French mercenaries in Mysore had learned to make them, and the British Ordnance Office was enquiring for somebody with expertise on the subject. In response, William Congreve, whose father was head of the laboratory at Woolwich Arsenal, undertook to design a rocket on Indian lines. After a successful demonstration, about 200 of his rockets were used by the British in an attack on Boulogne in 1806. Fired from over a kilometre away, they set fire to the town. After this success, rockets were adopted quite widely by European armies, though some commanders, notably the Duke of Wellington, frowned on such imprecise weapons, and they tended to drop out of use later in the century. What happened next, however, was typical of the whole British relationship with India. William Congreve set up a factory to manufacture

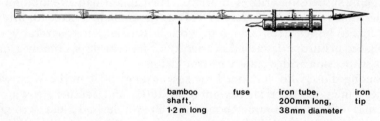

bamboo shaft, 1·2 m long   fuse   iron tube, 200 mm long, 38 mm diameter   iron tip

Figure 30   Indian military rocket of the 1780s or 1790s.
Such rockets weighed up to 4 kilograms each before firing and had a range of about 1,000 metres. They could cause chaos when used against cavalry by frightening the horses, and were also employed as siege weapons.
(Drawn from information given by Carman, Ray, and Needham, *Science and Civilization in China*, volume V, part 7, pp. 517–19.)

the weapons in 1817, and part of its output was exported to India to equip rocket troops operating there under British command.

Yet another aspect of Asian technology in which eighteenth-century Europeans were interested was the design of farm implements. Reports on seed drills and ploughs were sent to the British Board of Agriculture from India in 1795. A century earlier the Dutch had found much of interest in ploughs and winnowing machines of a Chinese type which they saw in Java. Then a Swedish party visiting Guangzhou (Canton) took a winnowing machine back home with them.[4] Indeed, several of these machines were imported into different parts of Europe, and similar devices for cleaning threshed grain were soon being made there. The inventor of one of them, Jonas Norberg, admitted that he got 'the initial idea' from three machines 'brought here from China', but had to create a new type because the Chinese machines 'do not suit our kinds of grain'. Similarly, the Dutch saw that the Chinese plough did not suit their type of soil, but it stimulated them to produce new designs with curved metal mould-boards in contrast to the less efficient flat wooden boards used in Europe hitherto.

In most of these cases, and especially with zinc smelting, rockets and winnowing machines, we have clear evidence of Europeans studying Asian technology in detail. With rockets and winnowers, though perhaps not with zinc, there was an element of imitation in the European inventions which followed. In other instances, however, the more usual course of technological dialogue between Europe and Asia was that European innovation was challenged by the quality or scale of Asian output, but took a different direction, as we have seen in many aspects of the textile industry. Sometimes, the dialogue was even more limited, and served mainly to give confidence in a technique that was already known. Such was the case with occasional references to China in the writings of engineers designing suspension bridges in Britain. The Chinese had a reputation for bridge construction, and before 1700 Peter the Great had asked for bridge-builders to be sent from China to work in Russia. Later, several books published in Europe described a variety of Chinese bridges, notably a long-span suspension bridge made with iron chains.[5]

Among those who developed the suspension bridge in the West were James Finley in America, beginning in 1801, and Samuel Brown and Thomas Telford in Britain. About 1814, Brown devised a flat, wrought-iron chain link which Telford later used to form the main structural chains in his suspension bridges. But beyond borrowing this specific technique, what Telford needed was evidence that the suspension principle was applicable to the problem he was then tackling. Finley's two longest bridges had spanned 74 and 93 metres, over the Merrimac and Schuylkill rivers in the eastern United States. Telford was aiming to span almost twice the larger distance with his 176-metre Menai Bridge. Experiments at a Shropshire ironworks gave confidence in the strength of the chains. But Telford may have looked for reassurance even further afield. One of his notebooks contains the reminder, 'Examine Chinese bridges.' It is clear from the wording which follows that he had seen a recent booklet advocating a 'bridge of chains', partly based on a Chinese example, to cross the Firth of Forth in Scotland.[6]

Chinese suspension bridges were simply constructed with chains made from wrought-iron links or bars (figure 31). The deck consisted of planks laid across several parallel chains, with other chains at hand-rail level. Thus the deck curved downwards to the centre of the span, providing an adequate bridge for pedestrians and pack-horses, but not for vehicles. There was nothing about such structures that Telford could possibly wish to copy. However, their long spans demonstrated the potential of the suspension principle. They were certainly a stimulus for the promoters of the proposed Forth Bridge, and there may have been a limited 'stimulus effect' for Telford also. As to more direct transfers of technology, it was America rather than China which made the more important contribution. James Finley's work in the United States was known to Brown and Telford from a book by Thomas Pope published in 1811.

The coming of the railroads

By the time Telford's great suspension bridge over the Menai Strait was completed in 1826, a new transport technology had emerged – the steam-operated railroad. Vital to this were improvements in iron-working which allowed strong rails to be made. The first steam railroad locomotive had been built by Richard Trevithick in 1802 at Coalbrookdale in Shropshire. His first convincing demonstration of a working locomotive followed two years later at Penydaren in South Wales. However, these locomotives were too heavy for the track on which they ran, and rails were frequently broken.

Around 1812, locomotives were introduced on a number of colliery railroads in northern England where horses had previously hauled wagons. They ran moderately well on cast-iron rails, but one major advance which made long-distance rail transport feasible came in 1820 or 1821, when a rolling mill at Bedlington ironworks in Northumberland produced strong *wrought*-iron rails in 4.6-metre lengths. These rails were laid on the Stockton and Darlington Railway which opened in 1825. Like most earlier railroads, this was built to serve coal mines, but passengers were also carried. Although horses were used to haul trains over parts of the line, locomotives had an unprecedented prominence.

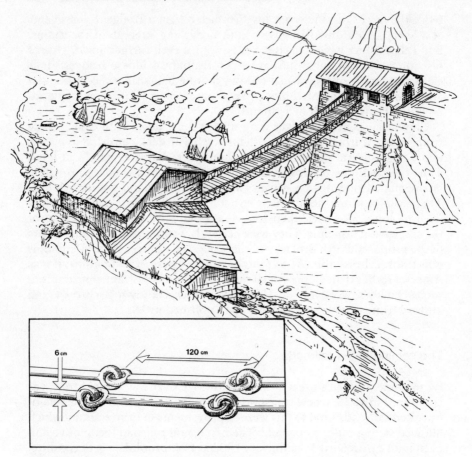

Figure 31 A Chinese iron-chain suspension bridge crossing the Niulan River, Yunnan Province. This was mentioned in an Imperial Encyclopedia in 1726, but is probably much older.

Of the two kinds of chain used in such bridges, one had conventional welded links about 30 cm long made from iron bars of 2 cm diameter. The other was formed of longer, heavier bars, as in this bridge. The inset shows the dimensions of the double chain at hand-rail level. There are other chains under the deck.

(Illustration of Jiang-di Bridge based on photographs by Leslie Pacey and Kenneth W. May.)

Many visitors came to see the line, including one from Pennsylvania, two Germans who were already advocating railroads in Prussia, and Marc Seguin, a French engineer who had built a suspension bridge in 1824 with wire cables instead of chains.

Ideas spread fast, and railroads were planned in many other places where slow and unreliable transport was an obstacle to trade. There were several elsewhere in Europe, while in the United States, the Baltimore and Ohio line was built between 1827 and 1830. It was laid out with full knowledge of the Stockton and Darlington Railway, and had the same type of rail and the same gauge. Meanwhile, between 1828 and 1830, George and Robert Stephenson built a sequence of locomotives, including their famous *Rocket* (a name made topical by a weapon still relatively novel in the West), in which they finally left Trevithick's locomotive concept behind and established the basis for a new generation of engines with tube boilers and horizontal cylinders.[7]

The first significant locomotive designer in the United States was John B. Jervis, who imported a Stephenson locomotive for a line he was building between Albany and Schenectady. Like all British locomotives at this time, this had only four wheels, and Jervis found that it lost power and in other ways performed badly on the sharp curves which were a feature of the line. After much discussion, Jervis designed a new locomotive with one pair of driving wheels at the back and a four-wheel truck or bogie carrying the weight at the front of the engine. This locomotive was built locally, and when it worked well, Jervis rebuilt the Stephenson locomotive in 1833 with a four-wheel bogie at the front. Its performance was greatly improved, and bogies soon became standard on American locomotives. Thus the steam locomotive in the United States was not simply the result of a transfer of technology from Britain. As in so many other examples, this was more like a dialogue in which there was an innovative response to the introduced technology.

One other invention which played a crucial part in the development of railroads as a transport system was the electric telegraph. Not only was the telegraph important in signalling the approach of trains, and hence in contributing to safety, but it was vital for efficient management of the railroad.

The electric telegraph became a practical possibility only when batteries of a reasonably reliable and economical kind were available. In England, such batteries were developed through researches by James Daniell at King's College in London, whose colleague Professor Wheatstone helped William Cooke develop the first British telegraph. Significantly, the first demonstration of this invention was on the London and Birmingham Railway between Euston station and Camden in 1837, and then on the Great Western Railway.

With this type of telegraph, signals were received by observing one or more moving needles which could point to left or right to indicate different letters of the alphabet. Even before this device was working in England, Samuel Morse had been experimenting with a different form of electric telegraph in America. His success again depended on the development of efficient batteries, with which he was helped by Leonard Gale, a professor from New York. Even more vital was the help provided by a mechanic named Alfred Vail. The crucial advantage of the Vail–Morse telegraph was initially that the receiving instrument marked the dots and dashes of the 'Morse code' they also invented onto a moving strip of paper. But an even greater advantage emerged when it was found that operators could decipher the sounds made by the instrument faster than they could read the message from the paper. Using a 'sounder' as a receiver, signals could then be taken down at 40 or 50 words per minute. In the course of time, this system became almost universal. Morse patented his equipment in 1840, and by 1850 telegraph networks were expanding rapidly.

Much has been written about the role of railroads in opening up the American west for settlement and agriculture after the first rail line crossed the Mississippi in 1856, but the characteristic pioneer railroads on the prairies had a line of telegraph poles alongside the track (figure 32). Railroad companies were prime users of the telegraph, and the two modes of communication expanded together.

Railroads and industrialization

Although the steam railroad was developed in Britain during the first three decades of the nineteenth century, its expansion after 1830 was faster in the United States. By 1840, some 4,600 kilometres of line had been built in America, compared with 2,400 in Britain and 1,500 on the continent of Europe. This rapid pace of construction in America was achieved by building cheaply, with sharper curves and steeper gradients than were usual in Britain.

As the rate of railroad building accelerated, unprecedented quantities of iron were required for rails and locomotives. In many ways, it was the technological success of the iron industry which made the railroad possible in Britain. Coke-fired blast furnaces provided the greater volume of production which was necessary. New methods for turning pig iron into wrought iron and rolling lengths of white-hot iron into rails then had to be devised before durable wrought-iron track could become commonplace (figure 33).

Given that these innovations had been made in Britain, it was inevitable that other countries wishing to build railroads at first imported the rails, if not the locomotives, from British manufacturers. At the same time, these countries improved their local iron industries rapidly with a view to making their own rails. In Europe, first Belgium and then Germany and France took this path.[8]

Figure 32   Railroads built across the American prairies in the 1860s were usually accompanied by telegraph lines.
(Illustration by Hazel Cotterell)

The situation in the United States was different since the industrial base was initially smaller than in France or Britain, but the need for railroads and the distances involved were greater. Thus the United States was soon taking a very high proportion of British exports of rails, and the prosperity of the British iron industry became closely linked with American railroad construction. Indeed, by supplying rails on favourable terms, some British iron-masters became important investers in American lines. It is instructive to observe that in the 1850s some 5,000 kilometres

Figure 33   Rolling wrought-iron bars and strips in a nineteenth-century ironworks.
Development of rolling mills such as this to produce wrought-iron rails was crucial if railroad track was to be manufactured in quantity and of adequate strength to carry the weight of a locomotive.
(From Edward H. Knight, *The Practical Dictionary of Mechanics*, London: Cassell, no date, by courtesy of Ironbridge Gorge Museum Trust.)

of railroad opened in Britain, but a staggering 30,000 kilometres were constructed in the United States.

Locomotives could be built by people with experience of making machines for other industries, notably for cotton spinning, and the workshops used had many tools in common. Thus, in the 1830s, locomotive building developed strongly in the machines shops near the textile mill complex at Lowell, Massachusetts, and at Philadelphia, Pennsylvania, where the Baldwin locomotive works grew out of an enterprise which had previously made machines for textile printing. The success of these and other engine builders was such that hardly any locomotives were imported after 1836, and in 1839 the majority of the 450 locomotives in the United States had been manufactured within the country.

The production of rails in any quantity was a more difficult proposition, however. It depended much more on bulk iron production, and hence on transformations in the iron industry which in America were just getting under way in 1830. Whilst blast furnaces for producing pig iron and reverberatory furnaces for converting it to wrought iron were broadly comparable to those in Britain, some redesign was needed so that anthracite coal, which was plentiful in Pennsylvania, could fuel them.

Progress was rapid, so that whilst 80 per cent of rails came from Britain during the 1840s, domestic production of iron rails overtook imports in 1856.

Through such developments, railroad construction stimulated the growth of ironworks and engineering workshops, not only in the United States but in several European countries also. The railroad and the industries which sustained it are therefore sometimes described as the 'leading sector' in mid-nineteenth-century development. On the continent of Europe, however, there was an important distinction between an industrial 'core' consisting of Belgium, Germany and northern France, and a 'periphery' in southern and eastern Europe. In the core region, much industrial growth can be convincingly traced to the impact of railroads. In the peripheral region, of which Spain can serve as an example, railroads were sometimes unprofitable and a burden on local institutions. Thus one historian[9] has noted that the 'railway dream', with its apparent promise of quick modernization, was so persuasive in Spain that for a while almost all available investment went into railroads, and 'true manufacturing companies' were neglected. Thus, although Spain acquired an extended railroad system, it did not develop industries whose raw materials and output would offer freight for the railroads to carry. Some of the most important lines ran to the ports, enabling Spanish ores to be shipped for processing in Britain by companies which paid handsome dividends to their British shareholders from the 1860s onwards, but contributed little to the Spanish economy.

By this time, railroads were being built in places even more peripheral to the European core, notably in Argentina. The first railroad there opened in 1857 with equipment imported from Britain. The first locomotive had been built for an Indian railroad, but when it was not wanted there, it was sent to Argentina instead. This meant that the track in Argentina was laid at the same broad gauge as in India, 1.67 metres. The episode seems typical of how arbitrary technical decisions could be imposed on 'peripheral' countries.

In other ways, Argentina was more successful than many other countries in using railroads to promote its national development. The government deliberately sought foreign capital to build new lines, but took steps to ensure that routes were chosen to meet local requirements, not just to produce profit for foreigners. If a line was needed through sparsely populated territory where little traffic could be expected, the government guaranteed the investors a 7-per-cent return on capital. By this means it was possible to push railroads into empty areas with good agricultural potential. The railroad thus became a means of settling some parts of the country, as in the United States, and gave ranchers and cereal growers access to markets and ports.

When railroad construction began to boom after 1880, lines were built very quickly, with some 8,500 kilometres of track laid between 1882 and 1892. Agriculture flourished, Argentina became a major exporter of grain to Europe, and some of the earliest refrigerated ships were tried out bringing cargoes of meat from there in 1876–7. By this time, Argentina's food-exporting industries were commercially successful and technically advanced, and large profits were being made. The money was not reinvested in further industrial development, however. Much of it returned to foreign investors. Argentine landowners who also became wealthy spent much of their cash outside the country. Another problem was that the food-exporting industry had become so profitable that it attracted the limited investment funds available to a disproportionate extent. Thus the economy did not diversify. The development of iron-works and engineering industries, which had been the medium through which railroads stimulated so much industrialization elsewhere, did not make progress.

This experience can be paralleled in many other countries, such as Mexico, where 20,000 kilometres of railroad were built between 1876 and 1910, allowing agricultural and mineral exports to increase, but with little real industrial development. The key question in every such instance is whether investment in railroads was balanced by investments in other industries. If it was not, then the railroads remained technologically dependent on the 'core' countries, and there was a continual flow of resources back from the periphery to the core. When we come to consider railroad development in Asia, we shall find that in Russia and Japan, large investments in supporting industries did take place. In China, a steelworks capable of manufacturing rails was opened before 1900, and a new coal mine in Hebei province, but hardly anything else. In India, where there was good potential for local investment in engineering and one steelworks did open, many obstacles were placed in the path of such developments.

Shipbuilding and Indian guns[10]

Before railroad construction is discussed any further, it is necessary to mention that steam power made its first significant appearance in Asia in the guise of the gunboat, not the locomotive. In 1824–5, the British discovered the potential of river steamers in warfare when they mounted an attack on Burma via the Irrawaddy River. They had three steamers in Indian waters at the time, and mobilized them for the operation, realizing that sailing ships would not be able to manoeuvre in the limited space offered by even as large a river as the Irrawaddy. Thus one task performed by the steamers was to tow warships up river. But they were also very useful in the fighting, and in the next two decades river-boats were deliberately designed with this role in mind.

An opportunity to use such vessels on a larger scale came in the first 'opium war' with China in 1840–2. The British had promoted the production of opium from poppies grown for the purpose in Bengal, and were exporting it to China in very large quantities in exchange for tea, which had a good market in Britain. The Chinese were alarmed by the scale of this drug trafficking. In a letter intended for Queen Victoria, one official commented: 'We have heard that in your honourable country, the people are not permitted to inhale the drug. If it is so regarded as deleterious, how can seeking profit by exposing others to its malific powers be reconciled with the decrees of Heaven?'

The British, however, saw the issue as one in which 'free trade' was more important than moral consistency. Thus Chinese efforts to stop the import of opium provoked 'punitive expeditions' in 1841–2, with British naval vessels proceeding up Chinese rivers. In 1842, the aim was to block the Grand Canal where it connects with the Chang Jiang River, because the importance of the canal in supplying grain to Beijing was well known. Several steamers were used, led by a heavily armed, iron-hulled steamboat named *Nemesis*, recently built in Britain by the Laird shipyard at Birkenhead.

In the 1850s, steam gunboats were used by Britain in a second opium war against China and a further war in Burma. Following the latter, a British administration was set up in the coastal zone of Burma, and this encouraged the formation of a private company to run commercial steamboat services on the Irrawaddy River. Rice exports expanded as transport improved, and more boats were built from parts shipped out from Scotland.

A different approach to the technology of steamboats developed from efforts to improve communication between India and Britain. The British community in Bombay favoured a mail route which would use a steamer from Bombay to the head of the Red Sea, with letters forwarded overland from there, and then via Mediterranean shipping. A 400-ton vessel was built for this service at the Bombay dockyard, whose earlier history was described in the previous chapter. It was launched in 1829 with two 80-horsepower engines sent out from Maudslay's works in London. The ship served its purpose well, but only after Aden had been taken by force to serve as a coaling station.

Experience with this ship stimulated one of the Bombay shipbuilders to make a detailed study of steam engines and other aspects of western engineering. He was Ardaseer Cursetji Wadia, who was mentioned earlier as building a gas-lighting plant and a foundry for making engine parts.

When the dockyard began to assemble steamboats on a regular basis about 1838, the East India Company appointed him Chief Inspector of Machinery.[11] The Company was collaborating with the Laird shipyard at Birkenhead, which had pioneered river-boats with iron hulls, and by 1841 Cursetji had supervised the assembly of five such boats from parts shipped out from Birkenhead.

Cursetji's work on engines may suggest ways in which an Indian response to western technology might have evolved, but in India's circumstances there was little chance for such 'dialogue' to flourish. With the development of screw propellers in place of paddles, as well as with the construction of iron hulls in place of timber ones, the technological gap between Europe and Asia increased. The screw propeller in particular opened the way for steamers to be as successful at sea as they were on rivers, and the first ship in which a screw was combined with an iron hull was SS *Great Britain*, completed to the design of I. K. Brunel in 1843. From about this time, the Peninsular and Oriental Company (P & O) grew to be the biggest operator of large steamers in Indian waters, and significantly, the only work they gave to the Bombay shipwrights was the construction of barges for use in coaling their vessels. The triumph of the steamer was completed in 1869 with the opening of the Suez Canal, which sailing ships could not use.

One branch of technology in which the Indian response to the West was very positive was the manufacture of guns. Indeed, technical progress in India after 1800 was the cause of many problems for British troops in the final stages of their conquest. The northwestern regions of the subcontinent were not taken until the Sikh Wars of the 1840s, and by that time Sikh forces were better trained than earlier Indian armies, and the local ruler had pursued an energetic rearmament campaign. New horse-artillery units were formed by the Sikhs in 1810, equipped with redesigned gun carriages, and skilled workmen were recruited in 1813 to staff newly extended arsenals. By 1821, foundries in Sikh territory at Lahore and Amritsar had produced 400 heavy guns and mortars, and before 1831 had turned out an additional 350. Muskets were also produced locally, but here there was a parallel policy of buying from abroad. This made it possible to take advantage of the latest western advances. For example, a chronic problem with matchlock and flintlock muskets was that they could not be reliably used in wet weather because the powder got damp. Chemical discoveries about other explosive substances, however, made possible the design of a more weatherproof gun with a 'percussion cap'. The Sikh army possessed such weapons in 1825, but it was not until 1836 that the British army approved its first percussion gun.

In 1843, the British advanced into Sind, and there were battles at

Hyderabad and Miani. The result was an intensification of military preparations in the Sikh heartland. The number of employees in Sikh arsenals and depots is said to have risen to over 50,000 in that year, and increased further to reach 72,000 in 1845. It is likely that too much was spent on arms and this led to 'fiscal breakdown'. Thus when the British finally occupied the territory in 1849, it was partly due to the economic exhaustion of the Sikhs, not to any inadequacy in their armament.[12]

American manufactures[13]

Firearms were now undergoing very rapid development in Europe, with the Minié rifle in France and the re-equipment of the Prussian army with breech-loading rifles. There was also a sequence of important innovations in the United States, including the Colt revolver, the Smith and Wesson repeating rifle and, during the Civil War, the first machine guns. These innovations considerably widened the power gap between the West and other nations, not just because the new weapons were more deadly, but also because their production depended on newly developed industrial techniques in *steel-making* and in applications of *machine tools*. Hitherto, guns had been made predominantly by craft methods both in Asia and Europe, and it was always possible for Asian craftworkers to copy western guns, or to produce designs of their own, often surpassing western standards of workmanship.

Steel production in the West was revolutionized by the introduction of processes which allowed this metal to be manufactured in bulk for the first time, and comparatively cheaply. Initially this was due to the Bessemer process, developed independently by William Kelly in Pennsylvania and Henry Bessemer in England, and patented by the latter in 1856. After early difficulties had been overcome, the Bessemer process was very extensively used in America.

Soon there was also the Siemens–Martin open-hearth process, and then Gilchrist Thomas and his cousin showed how steel could be made with ores which had hitherto been unusable in either the Bessemer or the Siemens–Martin process because of their phosphorus content. This brought down costs sufficiently for the new steels to become standard for army gun barrels. Thus, by 1875, European guns could no longer be copied or mended by smiths in non-western countries.

The other new approach to the manufacture of firearms originated in the United States during the 1820s, when the federal government set up an experimental factory for making rifles at Harpers Ferry, Virginia. The aim was to make components for guns to very precisely specified dimensions, so that any component would be *interchangeable* between one gun and another. Guns could then be repaired quickly by replacing damaged parts with spares taken from stock, rather than having to shape a new part specially, and assembly of new guns in the factory could be much faster. The federal Ordnance Department was soon requiring interchangeable parts in all firearms made for it, whether at its main Springfield Armory in Massachusetts, or by private firms. To this end, contractors were given access to the techniques developed at Harpers Ferry, which consisted of gauges and jigs used in conjunction with lathes and other machine tools to ensure that metal components and wooden gunstocks were all machined to precisely standard dimensions.

It was British visitors to the United States who found this approach so distinctive that they called it the 'American system of manufactures'. Its importance lay in the fact that the manufacture of intricate mechanisms for guns or other devices could now be organized in a manner analogous to the factory system which had developed earlier for textile manufactures. In other words, with fewer skilled workers, and with machine-paced work, a bigger quantity of guns, clocks, Yale locks (from 1855) or any other standardized mechanical product could be made than by reliance on craft methods. Sewing machines (after 1846), typewriters (from 1868), harvesters and bicycles were other machines made this way, and the widespread use of some of them would have been inconceivable without the ability to make standard, interchangeable components.

The principle sounds simple enough, but there were many problems involved in actually manufacturing components with such accuracy that they would fit together easily without the need for skilled workers to be constantly filing down oversized parts or making other adjustments. New machine tools were needed, such as the turret lathe invented in the 1840s and the universal milling machine of 1861, and methods of automatic control had to be devised so that these tools would finish machine parts with the right degree of precision.

Not only was this sequence of inventions important in itself, but it has been hailed as marking the point at which America ceased to be a net borrower of technology from other nations and became a key initiator of technological change. In the sense that the American system of manufactures represented a radically new direction for technical progress, this viewpoint clearly expresses some truth, but it does scant justice to the ingenuity of earlier American innovators, such as James Finley, Oliver Evans (pioneer of the high-pressure steam engine), John Jervis, and the co-workers of Samuel Morse. None of these were simply 'borrowing', nor were they reliant solely on technologies 'transferred' from elsewhere. Rather, they were using technical ideas which were part of an international discourse or dialogue to create entirely new inventions. James Finley was the first serious builder of iron suspension bridges outside China. Evans's engines and Morse's telegraph were not identical to contemporary British counterparts, but were in many respects more successful. Similarly, ideas about interchangeable parts and precision machine tools had also been part of an international discourse, having been applied earlier in Britain, albeit only to the manufacture of specialized products such as pulley-blocks for ships' rigging, and some textile machinery.

Few radically new inventions are made without some dependence on ideas already in circulation, and few transfers of technology involve a truly one-way transfer of machines or concepts. Technical progress is more often the result of a dialogue – of exchanges of technical ideas.

Indian mills and steelworks[14]

The role of several Indian techniques in this international dialogue has already been discussed, and the point was made that if shipyards had been allowed to develop, they might have become centres from which 'modern' skills could have spread into other Indian industries. After 1853, the same might have been said about railroad workshops.

Under British occupation, however, shipyards tended to contract, and when the construction of railroads began, Indian participation in any skilled role was excluded. India was forced to conform to the concept of 'transfer of technology' in its crudest form, in which one party is seen as a passive borrower of techniques, not a participant in dialogue.

The first Indian railroad was a short line out of Bombay which opened in 1853. Other railroad companies were soon formed under an arrangement by which profits accrued to the British investors, while losses were paid for out of Indian taxation. By 1870, there were 7,500 kilometres of route open, but after this date many lines were built by government rather than by private companies, often using the narrower metre gauge to save money.

By 1902, British India had a bigger length of railroad than all the rest of Asia put together, but there was little related industrial development. Coal mines had been opened to provide fuel for locomotives and steamships, and some coal was being exported by 1900. But when large railroad workshops were set up in Bombay, their chief function was to assemble locomotives and rolling stock from parts made in Britain. There was no growth of an engineering industry around them. Indeed, the only real engineering works in Bombay at the end of the century belonged to a manufacturer of cranes.

India was viewed by the British as an agricultural country, and land taxes were manipulated to encourage the production of export crops such as raw cotton and indigo. After the Suez Canal was opened, wheat and maize were also exported to Europe. Simkin wonders how a country impoverished by the decades of deindustrialization could afford to sell 10 per cent of its cereal crop, and implies that India's poverty was further worsened by this trade. However, British policy did lead to some useful investments in irrigation works, both to promote export crops and, after the famines of 1877 and the late 1890s, to prevent starvation.

To begin with, the British hydraulic works comprised little more than the reconstruction of old irrigation systems which had been allowed to decay. Among such works were the rebuilding of the Cauvery Dam in South India and the restoration of the Jumna Canals near Delhi. Later, much genuinely new construction was put in hand, opening up new land for cultivation in Sind (from 1873) and in the Punjab (between 1880 and 1901).

British investment in agriculture and railroads was not the whole story, however, for there were still some Indian merchants with capital to invest. They had survived the worst period of deindustrialization by trading on a small scale or acting as intermediaries between British firms and local producers of raw cotton, indigo and opium. Thus, although the British steered their investments away from anything that would promote industrial development in India, there were entrepreneurs who took a different view. Their opportunity came in the 1850s, when the cheap cloth from Lancashire which had put so many Indians out of business was beginning to rise in price.

In 1851, the first cotton mill equipped for mechanized spinning was opened by an Indian businessman. During the 1860s, conditions improved considerably because the Lancashire industry was in difficulties, with its supplies of raw cotton interrupted by the American Civil War. By 1870, there were 12 mechanized mills in India, mostly in Bombay and Ahmedabad, and they contained machinery equivalent to 0.27 million spindles. This number had increased to 1.6 million by 1887, and 5.0 million by 1900. Because of the lack of a local engineering industry, all this machinery had to be imported, and was sometimes dependent on European technicians for maintenance. At first the machinery all came from England, but local mill-owners did not stick slavishly to British technology and soon there were imports of more advanced ring-spinning machines of American design.

Jute was another Indian textile fibre, grown mainly in Bengal. From 1800 the East India Company had encouraged its export to Britain for making fabrics comparable to the coarser kinds of linen. Three decades later, Dundee flax spinners in Scotland were beginning to use their machinery to spin jute, and Dundee factories progressively imported more of the Bengali fibre to make sacking, hessian and canvas. For a long time, India functioned only as the raw material supplier, but eventually jute mills were established there. However, the first to be wholly owned by an Indian entrepreneur did not open until 1918. It was founded by G. D. Birla, whose previous experience was in trade — especially cotton – not in manufacturing.

One striking feature of these developments was that many of the entrepreneurs belonged to the same social and ethnic groups as the Surat merchants and Bombay shipwrights of the previous century. Many (though not Birla) were Parsis. The most prominent was Jamsetji Tata, who founded a cotton spinning company in 1868, and opened the locally famous Empress Mill at Nagpur in 1877. Others were from the Wadia family, whose role in shipbuilding (and steam engineering) was mentioned above. The Wadias established the Bombay Dyeing, Spinning and Textile Mills, which had 180,000 spindles.

A small modern iron industry had been started in Bengal in 1875, but there was very little growth of other heavy industry until Jamsetji Tata invested part of the fortune he had made from his cotton mills to found the Jamshedpur steelworks. The idea for this enterprise was his, and he chose the site and did much preliminary planning. However, he died three years before the steelworks opened in 1907. This venture actually received some encouragement from the British, who provided technical support and agreed that the Indian railroads would buy steel rails turned out by its rolling mill. It hardly need be added that Tata Industries subsequently became the largest industrial combine in India, with extensive motor-vehicles factories as well as engineering and metal-working enterprises.

It is difficult to avoid the impression that such developments came late in India because railroads were not allowed the catalytic role they played in Europe and the United States, nor could shipyards develop their potential. For a long time, investments in new rail lines were not accompanied by investments in engineering industries and iron furnaces. Steelworks and textile mills, when they came, did not transform the whole of Indian society as western societies had been transformed, but remained islands of 'advanced' development surrounded by a more slowly changing and more traditional agricultural economy.

9 Railroad empires, 1850–1940

Rails in Russia[1]

The first transcontinental railroad in the United States was completed with great publicity in 1869, linking the hitherto remote western settlements and the more densely populated eastern states. Further north, completion of the Canadian Pacific transcontinental line in 1885 demonstrated that Canada had effective government from coast to coast, and headed off ideas in the United States about a northwestward diversion of the border.

The lessons were not lost on the old empires of Asia, some of which similarly sought to use railroads to demonstrate sovereignty over remote territories and encourage economic and administrative development. Thus the so-called 'gunpowder empires' of the sixteenth and seventeenth centuries (chapter 5), where they still survived, began to turn themselves into 'railroad empires'. The governments of Russia, the Ottoman Empire, and eventually China saw railroads as a major instrument of modernization. Japan built many railroads in the process of extending its own empire, and of course, railroads in India were built to consolidate British rule.

In Russia, some railroads were constructed with similar empire-building purpose (as with the first line to Tashkent, which followed annexation of that region). Others, however, were related to policies for industrialization or for the settlement of hitherto empty territories and the development of their agricultural potential (notably in western Siberia). The first major Russian railroad was the line between St Petersburg and Moscow, begun in 1843 and opened in 1851. An American adviser was employed, but the czar insisted that the railroad should be the work of Russian engineers, and that the rails should come from Russian ironworks (with some imported from Britain). The former Alexandrovsk gun foundry near St Petersburg was re-equipped as a locomotive works with American help, and all the engines and rolling stock were built there. Many of the bridges were at first constructed in timber. The engineer, Zhuravskii, designed several with neat wooden arches whose stone abutments were

150

solid enough to support wrought-iron girders when the timber was eventually replaced (figure 34).

The construction of other lines soon got under way, but as the programme expanded, it was no longer possible to depend so fully on Russia's own resources. Many locomotives were now supplied by the Baldwin works in Philadelphia, Pennsylvania, and rails and investment funds came from France and other western countries. This led to much unhappy experience. Bridges built by foreign engineers were not designed to allow for the spring floods (of which the engineers had no experience) and stonework was often badly built. Thus there were many structural failures. At the same time, foreign companies often spent money extravagantly on non-essentials, then exploited the Russian government's guarantees of a return on capital to claim state funds.

By 1878, European Russia had a rail network over 20,000 kilometres in extent and the first line to cross the Urals into Asia had just opened.

Figure 34   Replacing a timber bridge on the St Petersburg–Moscow railroad in the 1870s.

This was a double-track line, so it was possible for one track to remain in use while spans on the other track were replaced. Here, the track still in use is carried by the original timber arch built before 1851, and a new wrought-iron girder, 16.7 metres long, is being lowered into place. The iron parts for this had been imported from Britain.

(Illustration from Ewing Matheson, *Works in Iron*, London, 1877, by courtesy of Ironbridge Gorge Museum Trust.)

An improved rolling mill, enabling more rails to be produced in Russia, had been set up, and from 1874 rails were being made of steel instead of wrought iron. An even more striking step toward technological self-sufficiency was the establishment in 1882 of a laboratory for research on locomotive design at Kiev – the first institution of its kind in the world. The steam locomotive may have been a 'transferred technology' originating in the West, but experiment and detailed innovation at Kiev under an engineer named Borodin led to a distinctive two-cylinder compound engine with excellent fuel economy. One version of this, a freight locomotive on eight coupled wheels, was so well suited to Russian conditions that over 9,000 engines of the type were built.

Meanwhile, Russia had acquired new territories in the eastern part of its empire, where the town of Vladivostok had been founded. Communication with this region was exceptionally difficult, but the completion of the Canadian Pacific railroad attracted much attention and helped gain acceptance for the idea that the whole 9,000-kilometre route from Moscow to Vladivostok should have a continuous line of rail. Construction of this, the Trans-Siberian Railroad, began at its eastern extremity in 1891, and in the west soon after. Four years later, agreement was reached on a 'short cut' to Vladivostok through Chinese territory in Manchuria, and a railroad opened here in 1903. The longer route which remained entirely within Russian territory opened in 1916.

Although this railroad was a state enterprise, others in Russia were built by private companies, many still with foreign investors. However, government industrial policies were now, in the 1890s, proving very successful. The worst practices of foreign interests had been curbed, and under a brilliant minister of finance, Sergius Witte, railroads were being developed in such a way that other heavy industries were encouraged. For example, in 1898 several steelworks were operating and there were 13 rolling mills producing rails. Engineering works and coal mines were also expanding. By 1913, Russia's industrial production was exceeded by only four other nations – the United States, Germany, Britain and France – and some 70,000 kilometres of railroad had been completed. But high industrial production did not necessarily indicate economic or military strength in such a vast territory. In 1905, Japan defeated Russia in a brief war fought partly at sea and partly on Chinese territory, and as a result took over the railroad in southern Manchuria which the Russians had built.

Japanese technology

The conventional wisdom is that Japan's industrial and military development was the result of a rapid transfer of technology from the West which began soon after the Meiji Restoration of 1868. This 'conservative revolution' ended Tokugawa rule (whose origin was mentioned in chapter 5), and gave Japan a more centralized government with the emperor as its focus. A modernizing civil service emerged, many of whose reforms were concerned with institutional changes in banking, company law and the land tax. The last of these produced revenues for government industrial initiatives, but meant that modernization was partly paid for by the continued poverty of peasant farmers. But despite the many reforms, there was also an important element of continuity, notably in the strong merchant class. For example, the Mitsui firm dated back to 1683. There was also continuity in technology, both with respect to the long-standing Japanese interest in western techniques, and in the persisting importance of traditional technology and local innovation.

It will be recalled that from the 1630s, Japan had limited its trade with the West to one small Dutch trading post. In 1720, the import of European books was liberalized. A number of scholars learned Dutch and studied medicine, agriculture and other technical subjects from the imported literature. Translations were authorized and circulated widely in this highly literate society. By 1820, there were schools of 'Dutch studies' or western knowledge, and a handful of young men were being quietly sent abroad to study western technology. The Tokugawa government of the period disapproved of such activities, which were encouraged by certain local rulers, notably in the provinces of Satsuma and Chosu. In these areas, ironworks of a kind entirely new in Japan were established soon after 1850. The Satsuma plant had two reverberatory furnaces for the manufacture of wrought iron, and machines capable of boring out large gun barrels.

The formal isolation of Japan ended in 1854 when eight American warships arrived to reinforce demands, made the previous year by Commodore Perry on behalf of the United States government, that certain ports should be opened for the provisioning of American ships. Perry expressed surprise as to how much the Japanese already knew about western science and engineering. The Japanese, for their part, were determined to construct ships that would be the equal of those from America. Thus the first steamship to be built in Japan was begun in 1862 at a government shipyard, and was completed with some help from France in 1866.

It should not be thought that early industrial development in Japan was all ironworks and steamships, however. Expansion of these industries was slow, partly because Japan lacked large deposits of iron ore, and partly also because of a shortage of capital. Japan's first steelworks did not open until 1896, two years after China's. Much more important during the 1870s and 1880s, as Okita and Lockwood point out,[2] were some of the traditional technologies, particularly in the textile industries and agriculture. Profits from these two sectors of the economy were extensively reinvested, largely via the land tax. Indeed, one of the earliest indications of the success of the modernization programme was a great increase in farm production, partly through an expansion in the area of cultivated land, but partly also through improvements in technique which led to a more intensive use of the land. This involved attention to transplanting rice, fertilization and irrigation. It allowed the growing population to be fed whilst at the same time generating resources from which other developments could be financed.

As to the textile industries, it can be strongly argued that one advantage Japan had over India was that it was able to make a direct transition from handicraft production to mechanized industry without an intervening period of deindustrialization. For example, one of Japan's most successful exports was raw silk. From 1868 to 1893, production steadily increased and the industry consistently provided more than half of Japan's export earnings. By 1905, Japanese raw silk exports equalled China's and supplied one-third of the world market. Soon after, Japan overtook China and became the biggest supplier to countries such as France which had large silk-weaving industries. Although improved silk-reeling machines played a part in this expansion, the more crucial changes were associated with improvements in organization rather than technique. There was better quality control through a government system of licensing producers of silk-worm eggs, and large-scale marketing of the product was undertaken by firms such as Mitsui.

In the cotton industry, large mills equipped with imported western *spinning machines* were introduced soon after the Meiji Restoration, but *weaving* continued to be organized in small workshops with semi-traditional, manually powered looms. Okita has analysed the effect of this in economizing scarce capital resources; it also meant that good use was made of existing skills in the work force.

Undoubtedly, the greatest of all Japanese industrial skills inherited from before the Meiji period were those relating to organization and commerce, and these skills were clearly demonstrated by the way in which the introduction of new technologies was managed. The example just given of imported spinning machinery used in conjunction with locally made looms is one example. Quality control in the silk industry is another. The employment of foreign technical advisers was also very carefully organized. In particular, all their costs were paid by the Japanese so that there would be no doubt about who was in control.

One of the largest groups of foreigners were the British railroad engineers and managers working under William Cargill, who was director of railways and telegraphs from 1871 in partnership with a local man, Inoue Masaru. The British supervised the building of the first railroad in Japan, the Tokyo–Yokohama line, which opened in 1872. But there was friction between government bureaucrats and the railroad administration, with Inoue Masaru threatening to resign. The eventual result was a decision to dispense with foreign assistance as quickly as possible. Thus the British departed abruptly during the construction of the Kyoto–Otsu line in 1881. After that, Japanese railroads continued to depend on imported rails and equipment (until after 1900), but foreign engineers were called in only when major bridges were designed.

Railroads and shipyards were encouraged particularly strongly by the government in Japan, and considerable subsidies went into these two industries between 1868 and 1913. Okita comments that railroads absorbed less capital than other forms of transport would have done, partly due to the adoption of the narrow gauge of 1.08 metres. He points to many benefits stemming from railroad development, but speaks of shipyards and heavy industry in terms of 'hasty construction . . . for military needs'. He notes that armaments production brought forth the machine-tool industry, although the manufacture of locomotives and textile machinery would also have required development in that direction.

Innovation and dialogue

When it comes to analysis of the transfer of technology into Japan, we need to note four channels through which the process operated: (1) the import of machines from the West, (2) the employment of foreign engineers (railroad, telegraph and machine-tool experts especially), (3) agreements with foreign companies (notably in connection with textile machinery and electricity supply) and (4) the rapid expansion of scientific and technical education. Here foreign teachers were employed, mainly at Tokyo University, which was founded in 1877. But Japanese teachers were so quickly trained that by 1893 there were no foreign professors left.

One of the foreigners at Tokyo University was a young Scotsman named Alfred Ewing who had worked on submarine telegraph cables as well as pursuing more academic studies at Edinburgh. After beginning work at Tokyo in 1878, one of his responses to local conditions was to invent an instrument for measuring earthquake vibrations. It was typical of the dominant interest of that period that the instrument was later used for measuring vibrations on a railroad bridge to check its safety. More significantly, Ewing and four Japanese co-workers carried out important researches on magnetic materials, which among other things had implications for the improvement of electric motors. After Ewing left at the end of his five-year contract, work of this kind continued in Japan, leading both to scientific advance and to innovations in alloy steels.

In the absence of a good and widely read history of Japanese technology, the assumption is often made that western techniques were simply being 'copied' at this time. The research on magnetic materials very clearly did not follow that pattern. The dialogue between Ewing and his Japanese colleagues initiated a line of research which continued independently in Japan until the 1920s and 1930s. In industry, imported machines initiated independent developments in textile machinery, electric lighting and locomotive design in a comparable way. Thus we may speak of a technological dialogue even where there was no personal contact, with imported techniques or devices stimulating local innovation.

In previous chapters, we observed this type of dialogue occurring during the transfer of early firearms technology from China to Europe around 1300 (which evoked the invention of the cannon in Europe), and the transfer of British locomotive technology to America about 1830 (which stimulated the invention of the bogie). A similar example from Japan concerns the situation in the 1870s and 1880s when modern spinning machines were being imported, but traditional methods of weaving were still in use. To cope with the potential for faster production provided by the new spinning equipment, it was necessary to improve the weaving process. One innovation made in Japan in response was a pedal-operated loom. Forty years later, when weaving had been mechanized also, the Japanese were importing automatic looms from the most prominent British manufacturer, Platt Brothers of Oldham. Very soon, however, Japanese technicians had come up with an improved automatic loom, the Toyoda. Now it was the turn of the British to learn from the dialogue, and in 1929 Platt's were given exclusive rights to manufacture the loom in England.

Innovation of this sort is often a reflection of social and economic conditions in the country at the receiving end of the technology transfer. For example, a relative shortage of labour is often said to have been a factor in America, where steam-powered mechanical shovels were introduced for railroad construction in the 1840s. This technology was not transferred back to Britain to any significant extent until the 1890s, however, because of a plentiful supply of badly paid unskilled labourers (many of them from Ireland).

The same kind of considerations apply equally to Japan. Population was growing fast, and there was great reluctance to accept capital investment or loans from overseas, so at the turn of the century labour was plentiful and capital was scarce. Weaving in small workshops and other traditional forms of production survived for a long time because they absorbed little capital. One economist speaks of many of the innovations which occurred as 'partial modernization' aimed at increased production without heavy investment. A somewhat different analysis identifies 'indigenous technical innovation' as one factor making capital investment go further, but also stresses the role of organizational innovations. For example, by reorganizing shift working, it was possible to get more production out of expensive machines.[3]

This again raises the question of the distinctive Japanese approach to organization. Technology can never be adequately understood in terms of machines and techniques alone. Machines are always used within a framework of organization and management, and often there are organizational changes at the heart of important technological developments. In chapter 6, we saw how the industrial revolution in the British textile industry originated with techniques from Italy and India, and long-standing traditional practices in Britain, all of them rethought and redeveloped to serve new insights into the organization of production. The result was the 'factory system' with its characteristic machine-paced work. When a mature history of technology is eventually written for Japan, we may be able to decide whether and how imported techniques were transformed to conform with different organizational concepts there.

Imperialist dimensions

The early industrialization of Japan coincided with the period when European imperialism was at its height, and when European nations were becoming increasingly competitive in developing armaments. In Britain between 1884 and 1914, expenditure on the navy increased three times faster than on the army, and naval engineering became the 'leading edge' of British technology, rather as railroads had been a leading sector fifty years earlier. Thus some of the ablest scientific minds were attracted to solving problems such as the accurate control of guns firing at very long range from moving ships. The focus of innovative work was located in the two biggest arms manufacturers, Armstrong and (from 1888) Vickers. Up to 1914, they were sharing technical information and patent rights with Krupp in Germany as well as with the leading French manufacturer. Thus, despite loyalties to national governments and competition as arms suppliers in foreign markets, technological development was a transnational enterprise.

Japan was drawn into this movement, though not as early as is sometimes suggested. There was no steelworks in the country capable of supporting such an effort until 1896, and the keel of the first large battleship to be built in Japan was not laid until 1905. Prior to that, warships had been bought from Britain, which also assisted in training naval personnel, and in 1902 the arrangement was formalized by the signing of the Anglo-Japanese naval treaty.

Japan had already flexed its military muscles, however, in a brief war with China in 1894–5. The issue was a conflict of interests in Korea, which had long been a Chinese dependency. When the Japanese defeated China, they expected to turn Korea into a colony, having learned by observation of the European powers how colonies could provide raw materials for industry and markets for manufactured goods. In the event, the western powers prevented a full Japanese take-over in Korea until 1910, but Taiwan became a colony, and Japanese goals in Korea were pursued by buying up its first railroad, which had been begun by American engineers in 1896.

When Britain had signed the naval treaty with Japan, its main concern was the rise of Russian power and the potential of Japan as a counterweight. Russia had not only built a railroad across Manchuria (part of China) to reach Vladivostok, but had leased the Chinese ports of Dalian and Lushun ('Port Arthur'). These had been connected by railroad with the Russian system, and considerable numbers of troops were employed guarding the lines. Fearing further Russian consolidation, the Japanese attacked 'Port Arthur' and in 1905 defeated the Russians both there and at sea. Under the treaty which followed, they took over the port and its railroad connections, and formed an organization known as the South Manchurian Railway to run them. This became the main agency for Japanese economic penetration of this Chinese territory, and was particularly important in giving access to iron ore reserves, shortage of which had been a considerable disadvantage to Japan hitherto. One judgement is that 'the South Manchurian Railway . . . deserves to rank with the East India Company and the Hudson's Bay Company as one of the great semi-governmental economic organizations of history.'[4] It was a classic example of the railroad as a means of operating extractive industries for an imperial power.

As for the rest of China,[5] after a disastrous period of civil war and rebellion (1852–60) and several invasions by European powers, there was much discussion of policies for recovery and 'self-strengthening' by the adoption of western technology. Progress had been made in the manufacture of modern weapons (from 1860), the operation of a profitable steamship line (founded in 1872) and the development of a telegraph company (1881). There was also some development of coal mines and an ironworks near Tangshan, and after 1890 there were cotton mills, soon with repair shops for servicing their machinery which played an important part in fostering new skills.

These enterprises depended heavily on engineers from Britain (on the steamships), Denmark (the telegraph) and other western countries. They were also limited by the way in which 'self-strengthening' policies were conceived only in terms of acquiring technology and technical education. There was no recognition of the need for new forms of organization to make the new technologies effective, and thus there were no institutional reforms like those in Japan after 1868. One result was a chronic difficulty in raising capital for new projects and arbitrary taxation of profits.

Other factors arose from the legacy of China's several defeats at the hands of Europeans after the first opium war of 1840–2. The European powers had demanded 'extra-territorial rights' for enclaves of traders and maintained naval forces at a number of 'treaty ports'. Whilst this enforced trade with Europeans did give some Chinese merchants experience of western trading methods, there were negative influences also. The development of railroads, in particular, was at first strongly resisted because all the proposed lines terminated at 'treaty ports' and were (correctly) seen as an attempt to extend European economic exploitation.

The first successful railroad in China was a short line built in 1882 to serve coal mines near Tangshan. Six years later, it was extended to the port of Tianjin, a distance of 145 kilometres. It seemed that there would be opposition to further extensions, but China's defeat by Japan in the war of 1894–5 encouraged thought about the need for more 'self-strengthening' development, and plans were soon being laid for a lengthy line running south from Beijing (Peking). After this, railroad construction proceeded fast. Finance came mainly from foreign loans, although there was some Chinese investment and management. At the same time, work was in hand at the Hangyang ironworks to install blast furnaces, an open-hearth steel plant and rolling mills for making rails. This was financed partly by Chinese shareholders, but there were also loans from a German company and a Japanese bank, and there was technical assistance from Germany. Output from the steelworks was at first disappointing, but it is significant that China could produce at least some rails for the construction of its own railroads from soon after 1894.

By 1913, the Russian and Japanese empires had substantial railroad systems. China had the beginnings of one, and as we saw in the previous chapter, India had a very dense network. In all four countries, there were modern steelworks and some other industry. What mattered for the future, however, was not the extent of these developments but who controlled them, and whether each nation was self-sufficient in the basic facilities necessary for running a railroad. This is a key point in the accompanying map showing the situation reached by 1940 (figure 35). Railroads are classified there as 'western-built' if constructed largely under western control to serve western interests. In Turkey, for example,

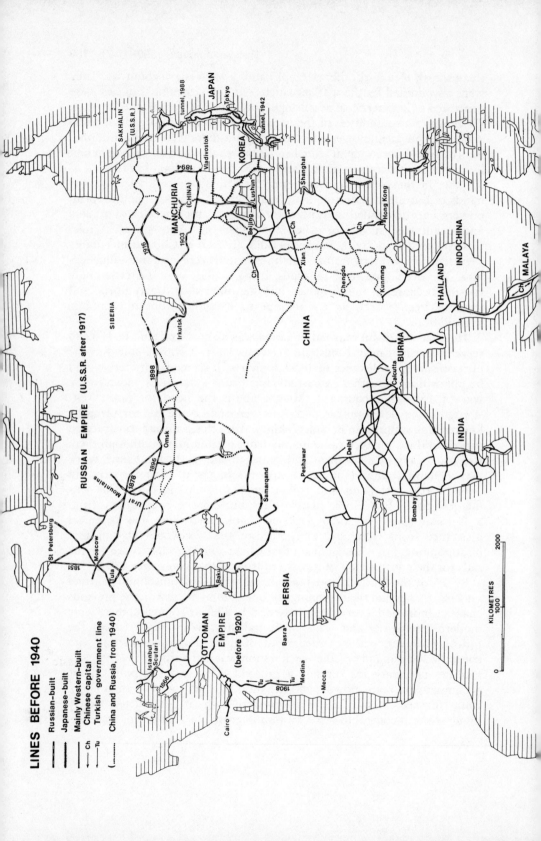

LINES BEFORE 1940

- Russian-built
- Japanese-built
- Mainly Western-built
- → Ch — Chinese capital
- → Tu — Turkish government line
- (········) China and Russia, from 1940)

RUSSIAN EMPIRE (U.S.S.R. after 1917)

SIBERIA

Ural Mountains

St Petersburg
Moscow
Tula
1851
1878
1896
1898
Omsk
Irkutsk
Samarqand
Baku
1916

SAKHALIN (U.S.S.R.)

tunnel, 1988

JAPAN
Tokyo
tunnel, 1942

1894
MANCHURIA (CHINA)
1903
Vladivostok
KOREA
Lushun
Beijing
Shanghai
Xian
Ch
Ch
Ch
Hong Kong

CHINA
Chengdu
Kunming

PERSIA

OTTOMAN EMPIRE (before 1920)
Istanbul
Scutari
1866
Cairo
Basra
Medina
Tu
Tu
1908
•Mecca

BURMA
Calcutta
Peshawar
Delhi
Bombay
INDIA

THAILAND
INDOCHINA
MALAYA
Ch

KILOMETRES
0   1000   2000

the concept of a line from the Bosphorus into Iraq and then to Basra was entirely the idea of a German engineer, Wilhelm von Pressel. It was built under German control and was left incomplete in 1914. Such lines were not adequately supported by related industries, but remained dependent on western supplies. By contrast, although railroads shown as 'Russian-built' were often constructed using western skills and investment, from the last two decades of the nineteenth century they were being built to serve Russian policies. They were also supported by rapid development of engineering industries, even before 1900. Japanese-built lines were almost wholly independent of western expertise and finance after the first decade of construction, even though rails and locomotives continued to be imported.

Although there has been a remarkable railroad-building programme in China since 1949 (see figure 35), slow progress earlier in the century is evident from the map. This was associated with China's internal institutional weaknesses, and was not due to any lack of business enterprise. Thus Chinese entrepreneurs operating outside the country were often more successful than those inside. Some of them, for example, bought up and operated tin mines in Malaya which European owners had abandoned as uneconomic. By 1913, Malayan tin output was 51,000 tons per year, three-quarters of it due to Chinese mining. This success seems to have been due to better judgement as to what technology would suit local conditions. The over-elaborate equipment used by the Europeans was commonly replaced by small and simple steam engines driving pumps with hydraulic jets for moving gravel. Chinese contractors also built railroads in Malaya, often investing capital where Europeans would not risk theirs. Here, as well as in China itself after 1914, it becomes

---

Figure 35 A railroad map of Asia, emphasizing lines built before 1940, and with dates of opening marked on selected routes.

The term 'western-built lines' indicates a great variety of arrangements between governments and entrepreneurs (local and foreign), also investors, engineers, contractors and suppliers based in Europe or the United States.

In the Ottoman Empire and China, foreign companies obtained concessions to build railroads which gave them great freedom of action. After 1905, there was a growing movement in China to gain local control, and after 1914 over 70 per cent of capital invested in China's railroads was Chinese-owned.

Russia and Japan took firm control of railroad-building in their empires after initial episodes in which there were uncomfortable relationships with foreign engineers, and in Russia with western companies. However, both countries imported rails and equipment from the West until after 1900.

(Sources include Chamberlain for Manchuria – strictly part of China – Issawi for Ottoman Turkey, Kumar and Desai for India, Liang for China, also Westwood for Russia.)

increasingly necessary to qualify the statement that local railroads were 'western-built'.

Silver geometry[6]

The completion of the first American transcontinental railroad in 1869 was widely portrayed in photographs and engravings showing a train from the Pacific coast meeting a train from the east at the place in Utah where the last rails were laid. The symbolism of this occasion, and the completion of other long-distance railroads, contributed to a powerful and optimistic vision which informed much technological activity. Submarine telegraph cables carrying messages under the oceans were part of the same vision, as could be seen from the enthusiasm with which people invested in the first Atlantic cable of 1857–8.

Needless to say, however, most railroads were constructed with practical and often very limited aims in view. In much of Africa and South America, the only lines built were links between mines or plantations and ports from which the raw materials produced could be shipped to Europe. Thus networks suitable for local travel hardly developed. In Brazil, for example, railroads to coffee plantations had tracks laid to different gauges, so even where their routes converged, trains could not run from one line to the other. Other railroads were built to open up internal markets for the sale of European goods, which was one motive for many lines in China and India. Some railroads were also planned to move troops and extend military control, notably in India and East Africa. Many such lines were built with narrow-gauge track to reduce construction costs, setting limits to the size and power of locomotives. Attempts to overcome this led to some striking innovations, such as the long, articulated Garratt locomotive, invented in Australia in 1907, and then manufactured by a British firm for railroads in Africa and India.

Sometimes, however, there was evidence of visionary planning inspired by the American transcontinental lines and the Atlantic telegraph. There was to be a Trans-Saharan railroad in Africa, and a Cape-to-Cairo line. One engineer in India wanted to lay tracks all the way from Europe to China. Later, the British built a line into the mountainous hinterland of Burma, and dreamed of a link with China's railroads. Near the head of this line was a big viaduct built for the British in 1899 by the Pennsylvania Steel Company which the travel writer Paul Theroux describes as 'a monster of silver geometry in all the rugged rock and jungle'.[7]

If that was one kind of symbol of imperial ambition, railroad structures within Britain itself were often conservative in design, with a grandeur appropriate to an imperial power in which the example of the ancient Roman Empire was remembered. This was explicit on the Great Western Railway where it approached the former Roman city of Bath through a tunnel whose entrances and other detail were deliberately Roman in design. A generation later, the Midland Railway was as ambitious as Brunel's Great Western, and there was a comparable symbolism in the heavy masonry of viaducts on its Carlisle line (figure 36), some again with Roman detail, as well as in the bolder monumentality of its enormous iron-and-glass station roofs in London (St Pancras) and Manchester.

However, the greatest of all European railroad engineers was Gustave Eiffel, who studied wind pressure on high bridges and took the mathematics of structural design further than most of his contemporaries. He built refined and adventurous viaducts in France and Portugal, and

Figure 36   Roman qualities in a British railroad viaduct.
  Arten Gill Viaduct, completed in 1875, on the Midland Railway line from Settle to Carlisle.
  (Illustration by John Nellist)

bridges in Indochina. But the culmination of his career was the application of the same iron lattice-work as he used for his bridges to build a work of deliberate symbolism but no utility at all, namely the Eiffel Tower of 1889.

Railroads had an imperial role for Japan as well as for the European powers. By 1931, the whole of Manchuria – part of North China – had come under Japanese control, and military and industrial developments there were based on the railroads. New lines were built to move troops to frontier areas. The Anshan steelworks, owned by the railroad company, became the largest in the Japanese Empire, with an output of 650,000 tons per annum. Soon there were plans for expansion to 3,000,000 tons capacity in the region, with a new railroad to a Korean port to provide a quicker route to Japan. The Manchurian ores were of low grade and difficult to smelt, but the Japanese had obtained rights to operate a German process which made smelting possible.

In 1937 Japan began a war with China, and as more Chinese territory was conquered, dreams of an ambitious 'railroad empire' became quite as visionary as any that British imperialists had entertained fifty years earlier. There were already plans for under-sea tunnels which would allow the three main Japanese islands to be covered by a unified rail network. The first of these allowed trains to run from Tokyo to Nagasaki on the south island of Kyushu, and opened in 1942. A 200-kilometre under-sea tunnel to Korea was discussed, and an Asian coastal line, extending Chinese and Indochinese railroads through Thailand and into Burma. Some parts of this line were actually built during the Second World War. Finally, in 1942, an Imperial Railroad Society was formed to promote the idea of a Berlin–Tokyo rail link which would pass north of Tibet, then through Iran and Turkey into Europe.

What is now surprising is not the fantasy of these schemes, but how near some of them have come to realization. In the 1980s, the second of the projected under-sea tunnels has been completed, and Japan is sending exports to Europe by rail – not by its dream line to Berlin, but via containers loaded onto Trans-Siberian trains after a short sea crossing. Meanwhile, other Asian states continued to develop their railroads, with a new Siberian railroad in the Soviet Union (the BAM line) and many new lines in China, and with railroad equipment now exported by India.

Production symbolism

In his discussion of the locomotive in America, the literary scholar Leo Marx makes the point that not only was it an important agent of change on that continent, but its public visibility made it the 'perfect symbol' of what was happening. People had only to see a locomotive to recognize that here was a new 'civilizing force', making settlement and development of the prairie possible. This encouraged a vision of a new world order somewhat more benign than the imperialist dreams of the Europeans. As the oceans were 'navigated by steam', the lands were traversed by rails and intelligence was 'communicated by electricity'. The poet Walt Whitman foresaw how the continents might be connected by a single, unifying network, 'cover'd all over with visible power and beauty'.[8]

If the railroad became a public symbol of what nineteenth-century technology was doing in the world, many other artefacts had a private symbolism for those who worked with them, and this also can tell us much about what was happening in this period. Sometimes, obscure parts of machines were ornamented in ways which show that they meant more to their builders than their mere utility would seem to imply. In Britain, for example, in any small water mill for grinding corn, every wheel and shaft was of solid and robust proportions and very occasionally there were decorative curves or mouldings, hidden away inside the machine where nobody but the miller would ordinarily see them (figure 37).

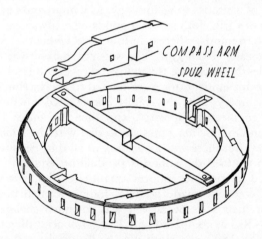

Figure 37   A large wooden gear-wheel (with teeth missing), known as a spur wheel, from a water mill at Black Bourton, Oxfordshire, England.

This demonstrates the very solid construction typical of eighteenth- and nineteenth-century mill mechanisms in England. However, craftsmanship is here taken to unusual lengths in the elaborate scarf jointing of the rim, and in the decorative curves carved into the spoke shown detached above the wheel.

(Reproduced by kind permission from *Oxfordshire Mills* by Wilfred Foreman, Chichester, West Sussex: Phillimore & Co., 1983.)

Similar detail can sometimes be found on other types of traditional equipment used for productive processes in Britain and Europe: windmills as well as water mills, spinning wheels and farm carts. Even after factory production was well established and machines were made wholly of iron, British looms, spinning frames and machine tools were solidly and proudly built, often with decorative flourishes.

It is very significant that in those parts of India and China which produced the finest quality textiles, European travellers frequently commented on the rough construction of looms and other equipment. Writing in 1873, a British traveller described a three-spindle spinning wheel he had photographed in China, emphasizing its ingenious mechanism, but noting the 'very primitive make' of its poor-quality timber construction and rough finish.[9] This was a reflection of the poverty of the textile workers, but it also indicated an attitude of mind which emphasized the quality of the final product without taking much interest in the production process.

In India, as was noted in a previous chapter, the highest-quality cloth might be laid out on the ground for painting or printing. Such practices reflected an economic situation where merchants found it more profitable to invest in trade than in improved methods of manufacture. But it was also indicative of attitudes to technology. Machines and equipment used for production of yarn, cloth or other goods were always of considerable interest to western minds, whereas in India and China they were perhaps only seen as means to an end. The western bias toward mechanical interests was clearly crucial for the industrial revolution in Europe, where so many important innovations were related to production equipment – first textile machinery, then the solidly constructed machine tools used in the British engineering industry, and then the much lighter, more precise lathes and milling machines associated with the American system of manufactures. But there was a paradox in all this, because even in the late nineteenth century the superior production equipment of Europe was being used to produce poorer quality textiles, paper and other products than could be made in Asia. Indian women bought British-made cloth for their saris because it was so much cheaper than cloth made in India – Rs3 as compared with Rs12. But according to one Indian report, written in 1890, the cloth was badly dyed, so that the colours ran when it was washed and a sari was unfit to be worn after four months.[10]

Similarly, paper made in Europe could not always equal the finest Chinese or Japanese product. Thus, in the 1870s, we find a major British publisher commissioning research on how to make a thin, opaque paper of comparable quality to some imported from the East.[11] The successful outcome of the research owed nothing to Asian technology, but the stimulus to do the research in the first place arose from the sort of 'dialogue' which has been referred to often in this book. Where quality of production is concerned, it is a dialogue in which western and Asian technology still have different approaches.

European interest in production equipment found its highest expression in attitudes to factory steam engines. The railroad locomotive appeared in public, drawing trains past platforms crowded with passengers, and it is understandable that its appearance was a matter of pride. By contrast the industrial steam engine operated inside a building, well away from casual public attention. Yet even in a Shropshire ironworks, when a large new engine was installed in 1801 to drive a rolling mill, its owner was said to be so pleased with it that he employed a woman especially 'to wash the engine house every day more than once and to keep the ironwork well blacked and everything clean'.[12] In British textile factories at this time steam engines were ornamented with decorative detail borrowed from Greek and Roman architecture. Later, more elaborate forms of decoration appeared, and the engine house at a factory or pumping station often looked like a small church. The engine was being presented as the most telling 'sign' of modern life. It was 'the controlling symbol for a new kind of culture'.[13]

10   Scientific revolutions and technical dreams

Electricity and chemistry

Until the 1870s, the development of applications of electricity was held back by the lack of a cheap source of electric current. Whilst there had been many experiments with electric arc lighting and electric motors, the only application of electricity which had come into worldwide use was the telegraph. This had been achieved because the telegraph used only very small and intermittent electric currents which could be supplied satisfactorily from batteries. The mechanical generation of electricity was expensive and inefficient because of the dependence of generators on permanent magnets. In 1866, it was discovered that these magnets could be replaced by coils of wire in which a magnetic field would develop by 'self-excitation' when the generator began to rotate. The first 'dynamo' based on this principle was manufactured by Siemens and Halske of Berlin in 1870, and an improved version designed by a Belgian named Gramme appeared in 1873.

The cheap generation of electric current opened the way to a much wider use of electricity in lighting, industry and transport. For example, a small electric tramway was demonstrated in Berlin in 1879. In the same year, Edison in America and Swan in England independently produced the first successful electric light bulbs with incandescent filaments. Within two years, Edison had small power stations supplying large numbers of his lights in both New York and London.

One feature of these innovations was the highly organized research work on which they depended, whether in Edison's workshop and laboratory in Menlo Park near New York, or at the Siemens and Halske works in Germany. Another notable point was the involvement of chemists such as Joseph Swan or, later, Eugene Obach, who worked for the Siemens firm. Chemistry was at the centre of a great many key developments in nineteenth-century technology, because better understanding of the chemical nature of *materials* was so often critical. For Swan and Obach, the materials at issue were those from which lamp filaments could be made. Elsewhere there were important discoveries about impurities in steel and means of dealing with them as a result of work by two chemists, Percy Gilchrist and his cousin, Sidney Gilchrist Thomas (see page 145). Most important of all was a long series of researches on coal and particularly on constituents of coal-tar, which was being produced in large quantities as waste from gasworks. These studies would soon transform coal-tar from an embarrassing by-product and pollutant into a valuable feedstock for the new organic chemical industry.

Although the foundations for this new science-based approach to industrial problems had been laid by French and British chemists, its fullest development occurred in the German-speaking countries of Europe, because it was there that *institutions* capable of sustaining the new science took shape most fully. For example, there were reforms in universities and new polytechnics were established, the latter firstly in the Austrian Empire at Prague and Vienna, then in the 1820s and 1830s at nine German cities, then later at Zurich. Some polytechnics were later reorganized as technical universities (*Technische Hochschulen*).

French influence, notably from the Paris *Ecole Polytechnique*, was important in all these developments. However, the German approach was often more practical. This was particularly true of the method of teaching chemistry which developed at the universities of Giessen under Justus von Liebig, and Heidelberg under R. W. Bunsen. At both places, teaching was closely linked to experimental research of a kind which could be, and from 1870 increasingly was, transplanted into German industry. The industrial research laboratory was the most important institutional innovation of the period.

The advantage which this gave to German technology became apparent when new dyestuffs were synthesized following research on the constituents of coal-tar. Paradoxically, the first of these dyes was actually made by an Englishman, W. H. Perkin, in 1856. Significantly, though, he was then a student of the German chemist A. W. Hofmann at the recently founded Royal College of Chemistry in London. In Germany, a big step forward came in the 1880s with the pioneer investigation by Emil Fischer of the molecular structures of sugars, proteins and eventually dyes. The chemical structure of indigo was determined in 1883 by A. von Baeyer, but it was not until 1897 that the Baeyer company's laboratories came up with a satisfactory method for synthesizing this dye. It could then be made more cheaply and in a purer form in a factory than if it were extracted from the indigo plant. In the year before the first production of the synthetic dye, Indian farmers had produced 9,000 tons of indigo worth about £4,000,000. By 1913, Indian production had fallen to 1,000 tons as a result of the new source of competition.[1] German indigo was being imported on a large scale by Japan, and even by India itself.

The largest firms which maintained laboratories were manufacturing dyestuffs and soda for the textile industries, and also fertilizers, paints, explosives and drugs (including phenacetin from 1887 and aspirin by 1899). For example, two chemical firms founded in the 1860s were Hoechst (located near Frankfurt) and BASF (at Ludwigshaven). By the mid-1880s, both had become large enterprises, each employing about 50 chemists in its laboratory. Baeyer and Company employed 15 chemists in 1881, but in 1890 built a new laboratory where there were soon 100 chemists with a supporting technical staff.

The electrical industry showed a similar development, not least because of the importance of chemistry in several of its branches. Apart from the work on lamp filaments already quoted, the improvement of batteries and electroplating stand out in this respect. Then at the end of the century new electrolytic processes for the manufacture of aluminium and chlorine came into industrial use. But electrical technology also depended on knowledge of mathematics and physics, subjects in which German universities and polytechnics also excelled. Thus it is not surprising that a number of key figures in the rapidly expanding American electrical industry had been trained in the German tradition. When Edison needed a mathematician to help design his first electricity supply systems, he employed Francis R. Upton, who had studied under H. L. F. von Helmholtz in Germany. Later, the development of alternating current supplies owed much to the work done for Westinghouse by Nikola Tesla, who had been trained in Prague, then in the Austrian Empire. This work did not really bear fruit until the 1890s, when the commercial generation of alternating current began to develop (figure 38). At the start of the next decade, when the General Electric Company founded a laboratory for basic research, we again find that several of its leading scientists had received at least part of their training in Germany.

The new engine

One other innovation which owes its origins in part to the high level of technical education and research in Germany was the internal combustion engine. As with industrial chemistry, however, we must note that some of the theoretical foundations had been laid in France, notably by Sadi Carnot. Another Frenchman, Etienne Lenoir, is noted for an experimental engine he built. In Germany, August Otto and Eugen Langen opened a factory in 1864 to make gas engines of their own design. These engines could use the public supply of coal gas as their fuel, and had a small, continuously burning flame to provide ignition. They suited owners of workshops who needed a smaller and cheaper source of power than the

Figure 38   Dynamo for generating alternating current.
This dates from the turn of the century, and consists of two sets of coils, one held stationary, the other rotating in close proximity.
(From W. Perrin Maycock, *Electric Lighting and Power Distribution*, London: Whitaker, 1904, reproduced by courtesy of Ironbridge Gorge Museum Trust.)

steam engine, and 5,000 were sold in a decade. By 1877, Otto had developed a more economical and much quieter four-stroke engine, but like the earlier one this depended for fuel on a gas supply. Meanwhile, the firm of Crossley Brothers had purchased rights to make Otto engines in Britain, and by 1900 had built many thousands. Often they were used by small businesses, and sometimes were fuelled by 'producer gas', made from coal by a small plant on the premises. They were also used to generate electricity for factories and institutions where there was no public electricity supply (figure 39).

Although Otto made experiments with benzene as a fuel for his engine, it was actually Daimler and Maybach who developed the first satisfactory engine to run on liquid fuel, in 1883. This ran much faster than the 150 revolutions per minute of the Otto engine, and the greater speed as well as the use of liquid fuel made it suitable for use in transport. Within two years, Gottlieb Daimler and, quite independently, Carl Benz had built pioneer automobiles.

Figure 39   Gas engine of the type invented by Otto driving a small dynamo (on the right).
   The engine has a single cylinder (on the left) and was made in England by Crossley Brothers.
   (From W. Perrin Maycock, *Electric Lighting and Power Distribution*, London: Whitaker, 1904, reproduced by courtesy of Ironbridge Gorge Museum Trust.)

These developments depended much more on empirical experience in the workshop than some of the innovations in chemistry and electricity mentioned earlier. However, theoretical knowledge of mechanical engineering as taught at the German polytechnics could often be significant. This point is particularly relevant to Rudolf Diesel, who traced the origin of the engine he invented to his training at the Munich polytechnic. In 1878, he heard a lecture there on Carnot's theorem concerning the ideal conditions for expansion of gases in an engine's cylinder. As he worked his way through the rest of his course and embarked on his professional career, this ideal, as he later wrote, 'pursued me incessantly'. Meanwhile his job entailed research work on refrigerators. Eventually, in 1893, with 'unutterable joy', he came up with a viable scheme for an engine that would approximate closely to the Carnot ideal, and he took out a patent and wrote a paper on the subject. The first 'diesel' engine was built by an Augsburg company soon after.[2]

Because so much of the history of technology has been conceived as the history of industry and written by economic historians, we are especially conscious of innovation which has been directed to economic purposes and whose success has depended on cutting costs and increasing the efficiency and profitability of production. Diesel's invention can readily be interpreted that way. In the lecture which gave him the urge to think about this subject, his teacher had said that steam engines transform 'only 6–10 per cent of the heat value of their fuel' into useful work! One could thus gain the impression that Diesel was motivated by an urge to save fuel and cut costs. Certainly we would never have heard of the diesel engine had it not offered sufficient economic advantage for first an Augsburg firm, and then the Sulzer company, to take up its development. But to suggest that market considerations were at the core of Rudolf Diesel's thinking would be to misunderstand how creative thinking occurs in technology. Diesel himself wrote that the concept of an ideal engine cycle had obsessed him – 'dominated my being' – through fourteen years of other distractions before it crystallized into an idea for a practical engine. That is the language of a man whose imagination is caught up with a dream, not a man who has merely recognized a market opportunity.

The swallows fly laughing

We can gain more insight into how dreams and technical ideals have been of importance in the twentieth century by considering one of the most important inventions of its first decade, namely aircraft powered by benzene or petrol engines.

Here the commanding ideal was not a theoretical concept, such as isothermal expansion in an engine, but the everyday observation of birds in flight. Since this had fascinated people throughout the ages, the ancestry of the aircraft is conventionally reckoned to include all sorts of vain attempts to fly, including the Greek legend of Icarus, stories about a medieval monk who made himself wings, speculative drawings by Leonardo da Vinci, and the remarkable series of gliders made by Sir George Cayley in England from the 1790s until 1852.

Cayley's work deserves more recognition, but after this, serious scientific work on winged devices was most fully developed in the work of Otto von Lilienthal, another German-trained engineer. As a young man he had experimented with models based on birds' wings, and later he returned to the subject, publishing a book on bird flight and wing structures in 1889. In 1891, he turned to experiments with what would now be called hang-gliders, in which he launched himself into flight from a small hill near Berlin. His flights were limited to distances of 100 to 250 metres, but he slowly learned how to control his gliders and experimented with flaps and elevators designed for this purpose. He died in 1896 after one of his gliders crashed, but this was after many flights over a period of about five years.

Lilienthal's influence was very great, partly because his technical writings were translated and widely read, but also because good photographs were taken of some of his flights. These pictures persuaded people of the need to understand gliding and to improve the design and control of gliders before powered flight could be achieved.

Among those influenced and inspired in this way was the Scottish engineer Percy Pilcher, who built his first glider in 1895, but visited Lilienthal before trying it. In America, Orville and Wilbur Wright praised Lilienthal's work very highly and used his drawings of birds' wings and gliders in designing their aircraft. In 1903, the Wright brothers succeeded in making the first short powered flight, using a biplane powered by a small benzene-fueled motor driving a propellor.

In none of this work were commercial motives very strong, and it would have been difficult to foresee the immense economic importance that aircraft were later to have. Several people with military interests or experience were involved, however, including Lilienthal himself and Hiram S. Maxim, the inventor of a machine gun. Balloons had already been used for military reconnaissance (and even for dropping bombs, at Venice in 1849), and the potential of aircraft in these roles must have attracted attention. However, Lilienthal was also a dreamer, pursuing an interest which had been with him since youth, and he expressed his dreams quite clearly in his writing. The 'longing' to fly gave him no rest, he said: 'a single great bird, circling above our heads, arouses in us the desire to soar through the skies as he is doing.' But as soon as men try to apply their knowledge in practical attempts to fly, 'our clumsiness is deplorably obvious, and the swallows fly over our heads laughing.'

It is perhaps not surprising that the man who turned from railroad engineering to build the Eiffel Tower should later take up aircraft design. Railroads were the subject of many quite explicit dreams of conquest, and the Eiffel Tower took the enterprise into a third dimension. Gustave Eiffel was already interested in aerodynamic forces and wind pressures on his high railroad bridges, and the tower posed the same problem in a more acute form. In 1906, he built a pioneer wind tunnel at the foot of the tower and used it in many experiments on aircraft wings. This work gave the first really clear understanding of how airflow around a wing could generate 'lift', and Eiffel went on to test models of many aircraft, and to develop a mathematical theory of propeller design.

Eiffel regarded the tower as his greatest work. It formed part of the Paris International Exhibition of 1889, and has been described as 'the man-to-the-moon program of the day, carrying technical innovation to a logical absurdity', in order to symbolize French national achievement and a sense of 'industrial Eden'.[3] Whilst a real man-to-the-moon project was still eighty years away, conceptually it was much nearer. Even during the first scientific revolution three centuries earlier, as people were groping towards an understanding of the moon's motion in the heavens, the theoretical possibility of reaching the moon became a dream, explicitly stated by Kepler in 1609–10. Once aircraft were flying it became logical to think again of other forms of flight, and the rockets used by European armies in the first half of the nineteenth century were an obvious model. In Germany as early as 1923, Hermann Oberth published a book on rockets in space. This inspired the formation of a society for the promotion of space travel in 1927. Two years later this society was joined by a very keen new member – a teenager named Werner von Braun.[4]

Von Braun was another dreamer, and chose a training in engineering subjects that would give him a technical background suitable for investigating the possibilities of space travel. He was an active participant in the space society's very serious experimental work on rockets, and when these were taken over by the German army in 1932 he chose to work for the army, because building weapons seemed to be the only available 'stepping stone' into space. As is well known, the culmination of the army programme was the V-2 rocket used to bombard British cities toward the end of the Second World War. After the first successful flight of a V-2 in 1942, von Braun seemed to forget the war completely and enthused: 'We have invaded space with our rockets for the first time . . . we have proved rocket propulsion practicable for space travel.' In 1969, it was rockets designed by von Braun and his team, now working in America, which made possible the first landing of men on the moon.

Dreams of new worlds[5]

The pursuit of technical ideals or dreams is readily apparent in the innovations associated with Diesel, Lilienthal and von Braun. Economic advantage and even military technology were not central to their motivation. By contrast, the work of chemists who developed new dyestuffs was often very close to the point of industrial application. It fits better with the pattern of market-led innovation assumed by most economic historians. Yet the chemists had dreams too, and sometimes let slip comments about 'exploring new worlds'. By this they meant that they were steadily learning more about the architecture of molecules and how this was related to the observed chemical behaviour of materials.

Sometimes also such words indicated speculation about the nature of the atoms from which molecules were built.

One challenge to chemists interested in molecular structure was the unusual and useful properties of natural materials such as silk or rubber. The idea gained ground that such materials were composed of very large molecules. Cellulose was identified as the substance whose large molecules were of most importance in a wide range of vegetable materials, including cotton and timber. Experiments based on the reaction of cellulose with various acids contributed to Swan's researches on lamp filaments in the 1870s. They also led to the discovery of a new explosive (cellulose nitrate or 'gun cotton'), a new transparent material (celluloid, first made in America in 1868) and a cheap substitute for silk (rayon).

These materials were all made by manipulating the large molecules of cellulose provided by nature. The first plastic material for which the necessary large molecules were synthesized was phenolic resin or 'bakelite', first made by Leo Baekeland, a Belgian, in 1907. More new materials were developed in the 1920s, including synthetic rubber and plastics based on thio-urea and urea formaldehyde. Whereas bakelite was always a dark-coloured material because of substances added to give it strength, the new plastics could be produced in light colours, often with a marbled effect. It was a sign of the times that some areas of China which had long exported porcelain dishes to Europe were now (in 1930) a market for plastic rice-bowls made in Germany.

Despite the industrial successes of organic chemistry, there was a continuing debate throughout the nineteenth century about the nature of the basic atoms from which molecules were made. Doubts about whether the chemists' atom was really fundamental were reinforced when a much smaller particle, the electron, was recognized in the 1890s. The discovery emerged from a long series of researches on electrical phenomena in which electrical discharges had been studied and radio transmission had been demonstrated (initially by Heinrich Hertz in 1887). Once the electron had been recognized, further work led to the invention in 1904 of the first electronic valve. The man responsible was J. A. Fleming, a consultant to the Marconi Wireless Telegraph Company, who was attempting to develop improved methods of detecting radio signals. Very shortly after Fleming's 'diode' valve came into service, Lee de Forest in America invented the 'triode' as a device for amplifying radio signals. These inventions quickly emancipated radio from reliance on the crude spark gap transmitters, mechanical alternators, and carbon-particle detectors which had characterized its first decade.

Another challenge to the nineteenth-century view of the atom came from the discovery of the elements radium and polonium by the Polish-born chemist Marie Sklodowska, later famous as Marie Curie, helped by her husband Pierre until his death in 1906. These substances were 'radioactive', as Marie Curie put it, and she was able to show that their emissions of radiation were associated with changes in the masses of atoms and in chemical properties also. If atoms could change like this, they could hardly be the ultimate particles of matter.

There had been speculation about the possible atomic structure of matter among the ancient Greeks, and intermittently ever since. The possibility of penetrating the secrets of these hypothetical particles had been a dream among philosophically inclined people for almost as long as the wish to fly. Now, at last, evidence seemed to be accumulating which really shed light on the matter, pointing to the existence of particles smaller than the chemists' atoms. Here, then, was another new world to explore, more challenging even than the world of molecular architecture. By the 1930s, there was intensive research in several European centres, notably once again in Germany, but also including Cambridge in England, where Ernest Rutherford led a very effective team. In Paris, Marie Curie's daughter Irène was investigating the effects of bombarding a small amount of uranium with a stream of the recently discovered particles known as 'neutrons'.

Similar researches were under way in Berlin, but there was considerable disagreement between the scientists there and Irène Joliot-Curie's group in Paris. Thus, when Irène's results were published, Otto Hahn in Berlin was dismissive of what he called 'our lady friend's writings'. In 1938, however, a colleague recognized the importance of a new paper from the Paris laboratory and pressed its findings on Hahn. Suddenly, Hahn saw the point and began an intensive series of experiments to check it out. He came to the conclusion that Irène Joliot-Curie and her co-worker (a Jugoslav named Savitch) were right about the materials produced when uranium was bombarded with neutrons. He saw that the nucleus of the uranium atom had split into large pieces, a process which became known as 'fission'. Other investigators soon discovered that neutrons were emitted during fission of uranium atoms which could potentially split other atoms, releasing more neutrons and initiating a 'chain reaction'.

Thus on the eve of the Second World War the door was opened to a new technology, and a new weapon.

Atoms for peace

In the 1920s and early 1930s, nuclear research was pursued innocently and idealistically as the exploration of an invisible world. It seemed the fulfilment of an old dream about understanding the ultimate structure of matter. As soon as the practical and destructive uses of nuclear discoveries began to be exploited, however, the nature of the research changed radically. Instead of being open and international, with the Cambridge laboratory, for example, playing host to Russian, Indian and Japanese scientists, research became secretive and bureaucratic. Idealism was replaced by the pursuit of vested interests as the 'atomic bomb' became the great symbol of national status and military power. The United States used its first nuclear bombs against Japan in August 1945. Soviet Russia tested its first nuclear weapon in 1949 and Britain exploded one in 1952.

For nuclear research to recover some of the idealism, internationalism and esteem it had previously enjoyed, there was a need to show that it could contribute constructively to human development. At the United Nations in December 1953, the American president, Dwight D. Eisenhower, took the initiative. He suggested taking 'this weapon out of the hands of soldiers' and giving it to those who could 'adapt it to the arts of peace'.

In discussions which followed over the next few months, there was much talk of nuclear devices 'in the fight against cancer'. There were also claims about how 'atomic rays will cut lumber', and it was said that an 'atomic' locomotive had been designed for the railroads. The scientists were dreaming again, and the public imagination was captured by the prospect of electricity 'too cheap to meter'. As one historical account had described it,[6] Eisenhower's phrase 'atoms for peace' became 'witchcraft', in the sense that it was used as a symbol for 'driving off' fear of nuclear war. In more practical terms, the new approach opened the door to the spread of civil nuclear technology, and soon a whole galaxy of nations had set up atomic energy commissions on the American model – including Germany and Japan, Spain and Brazil. Japan signed agreements on nuclear co-operation with the United States and Britain and arranged to buy its first nuclear power station from the latter. Meanwhile, the first of a large number of Indian scientists came to America for training in nuclear physics.

India was the first Asian country to launch a nuclear programme, which was not entirely surprising in view of its established tradition of participation in western science. In 1895, for example, Jagadis Chandra Bose had made major improvements in the carbon particle 'coherer' then used in radio receivers and employed it for studies of very short radio waves. Another Indian scientist of the same name is remembered as co-author with Einstein of important work in quantum mechanics, the Bose–Einstein statistics of 1924–5. The pioneer of Indian nuclear research was Homi Bhabha, who had studied science in England, and who came from the small group of families in Bombay which had produced famous shipbuilders in the eighteenth century as well as nineteenth-century industrialists such as Jamsetji Tata. Bhabba's family was related by marriage to the Tatas, who had founded and endowed a university institute for the study of science in 1911. Now, in 1944, Tata Industries established the Institute for Fundamental Research to ensure that nuclear physics would have a home in India, and that Bhabba's genius would not be lost to the nation. An atomic energy commission was set up by the government in 1948, soon after India achieved independence, with Bhabba as its first chairman. In 1954 funds were made available for the construction of a research reactor.

The Indian government's decision to invest large sums in nuclear energy has to be understood in terms of the scars left by the long period when Indian industries had been deliberately run down by the British to ensure that there was no Indian competition for their own products. (See chapter 7.) This period of deindustrialization had not only impoverished the country and wasted skills, but had undermined confidence among Indians in their technological abilities. Even when it became possible for Indian-owned industries to develop on a large scale, British control of the national economy appeared to set limits on what they could achieve. What Bhabba believed, and what Prime Minister Nehru quickly accepted, was that an Indian success with a modern technology would give Indian scientists and engineers self-confidence in the tasks of national reconstruction which lay ahead.

The idea that India needed a technological success to rebuild confidence was probably well founded, but the choice of nuclear technology for this symbolic role proved most unfortunate. By the mid-1970s, nuclear power programmes in most western countries were in trouble, with orders for power stations cancelled or delayed because of their unfavourable economic performance, and because problems of safety and waste disposal were unresolved. Bhabba's plans for India had been particularly ambitious, in that he envisaged a fuel cycle based on thorium rather than the more usual nuclear fuel, uranium, since India had large thorium reserves. The plan was to begin with a series of uranium-fuelled power stations. One was bought from the American firm General Electric and others were begun with Canadian collaboration. Plutonium produced in these stations was then to be used with thorium in specially designed breeder reactors, but this part of the plan became a steadily more distant prospect.

Had India's nuclear programme fulfilled its initial promise of providing cheap and reliable electricity, it could have contributed greatly to the country's prosperity. Most electricity production from all types of power station was used by industry, but a surprisingly large amount was used in agriculture. Irrigation water, lifted from wells by electric pumps, was one of the key resources for the big increases in grain production associated with the so-called 'green revolution'. The number of electric pump-sets increased eight-fold during the 1960s, and passed two million in 1972. By this date demand for electricity was outstripping supply, and in order to keep the irrigation pumps running, industry suffered electricity cuts which led to serious dislocation and a decline in production. There were delays in construction and maintenance of all types of power station, but by the end of the 1970s the nuclear stations had the worst record of all. The four nuclear stations open at the start of the next decade were generating only 600 megawatts on average, well below the 2,700 megawatts for which they were designed. Only 2 per cent of total electricity output in India came from nuclear sources.

One interpretation of India's nuclear programme is that it was a 'national disaster', undermining industrial development and diverting investment away from low-cost hydroelectric projects. But that is an openly partisan verdict. An earlier and more carefully worded report for the World Bank comments that 'as in most other countries . . . performance of nuclear power stations in India has fallen below expectation.' But some people were still speaking of 'Asia's nuclear future'. Some also seemed to think that for India to retreat to supposedly more 'appropriate' technology would entail another loss of confidence like that associated with deindustrialization. It would lead to 'colonization' of Indian minds and sciences.[7]

By 1980, Japan had the largest nuclear power programme in Asia, but anyone who looked there for encouragement would not have been totally reassured. Bitter experience with a first nuclear station bought from Britain was followed by better results from the next plant, based on American practice. But even in the 1980s, with over 30 nuclear stations generating a quarter of the nation's electricity, there were worrying problems of reliability. These were of particular concern in view of Japan's very limited coal and oil resources.

China's approach to nuclear technology focused entirely on the military aspect. After a century of intermittent warfare, including several invasions by European powers before the revolution of 1911 and the long war with Japan which ended in 1945, it was understandable that China should want to acquire a weapon which seemed likely to deter all further aggression. By 1958, Soviet Russia was helping China construct a uranium enrichment plant. But Russia refused to provide a sample bomb, and when Soviet advisers were suddenly withdrawn in 1960 following political disagreement between the two countries, the enrichment plant was incomplete. However, the Chinese successfully brought it into production and used the uranium so obtained to make nuclear weapons, the first of which was tested in 1964.

Two legs in China and Japan

It may appear that with technologies such as nuclear energy, or electronics – or even with the railroad in the nineteenth century – the package of engineering products and techniques developed in the West is so complete, and so locked in to western science, that the possibility of a technological dialogue with non-western cultures is almost nil. Countries wishing to use these products and techniques would seem to have no choice but to accept the transfer of ready-made technologies. National policies may, of course, influence the way certain technologies are used, as when India aimed at a thorium-based fuel cycle and China chose a weapons-first approach to nuclear technology. There can also be elements in national cultures which affect the way in which a technology is used, such as the initial difficulty of producing a computer output in the Chinese language (solved for one type of computer by the Stone Company of Beijing in 1984), or of devising a Chinese word processor (in production from 1986). However, these are all relatively superficial modifications in what are essentially western-style technologies, and the potential for technological dialogue would seem to be very limited.

A different view emerges, though, if we take a broader view of technological development. Many of China's policies prior to 1976 were encapsulated in the phrase 'walking on two legs'. This had many meanings, some now discredited, but an important part of it was the idea of developing industry by building large- and small-scale installations simultaneously. Small fertilizer factories or hydroelectric power stations were often better suited to rural areas and could be completed more quickly than large ones, an advantage outweighing their poorer operating efficiency. Another aspect of this policy was the good use made of local skills. For example, while eight large fertilizer factories built before 1975 were assembled from imported equipment, all the hardware for many hundreds of small factories was made in China, much of it in Shanghai. Some, indeed, came from an engineering works there which had been established in 1902 to repair machinery for steamships and textile mills. This had expanded over the years, and by the 1960s and early 1970s it provided a reservoir of very experienced workers, capable of turning designs for small plant into working equipment. Construction and operation of the fertilizer factories in rural areas not only depended on this equipment but also drew on the skills of many local people.

In some instances, 'walking on two legs' would involve people skilled in older forms of technology, so initiating a 'technological dialogue' between western and traditional techniques. This occurred particularly in silk and other textile industries, but a more striking and less rational example was the rapid expansion of iron production about 1958 based on the revival of traditional types of small-scale blast furnace, suitable for 'backyard' operation. Since a few of these furnaces had still been in use in the 1930s and 1940s, some of the necessary skills were available, and for a time large amounts of iron were produced. However, enormous quantities of fuel were used (much of it charcoal) and the output was generally of poor quality. Thus the experiment was soon discontinued and is now dismissed as one of the follies of the 'Great Leap Forward'. However, useful lessons were learned about the operation of unusual types of furnace.[8]

In some aspects of engineering, and also medicine, 'walking on two legs' led to more fruitful interactions between modern and traditional technologies. For example, a number of arched bridges built of stone or concrete were based on the principle that a very flat arch of considerable span can be built of stone provided that excessive weight in the haunches of the arch is avoided. The classical example is Zhaozhou Bridge in Hebei Province, which survives from AD 617. Here the necessary weight reduction was achieved by building the haunches as a series of subsidiary arches instead of using solid masonry. During the 1960s, this type of bridge became standard wherever arched bridges of masonry or concrete were built. Some examples were constructed during the building of the Chengdu–Kunming railroad, notably the 54-metre span Strip-of-Sky Bridge (figure 40). One account of this structure notes its similarity to the historic Zhaozhou Bridge as an illustration of the maxim, 'Make the past serve the present.'[9] But there was never any idea of relying totally on traditional techniques. Concrete beams of standardized design were used extensively for viaducts on this railroad, and when the Zhaozhou concept was applied elsewhere, many inventions aimed at further weight reduction were introduced, especially with concrete construction.

Turning to a very different field, we may note that in medicine Chinese achievements over the same period were sufficiently impressive to attract a semi-official delegation from the United States in 1974. The delegates reported that Chinese medical practice was too deeply rooted in Chinese culture for any direct 'transfer of technology' to America. However, the report documented innovations in traditional Chinese practices (which include acupuncture and herbal therapies) and noted that in medicine, 'walking on two legs' meant that western-style medicine, surgery and biochemical research were practised side by side with traditional medicine. The biochemical work is worth mentioning because it illustrates the continuing importance of investigations of molecular structures for some technological achievements. In this case, the achievement was that of Chinese scientists being the first anywhere in the world to synthesize the hormone insulin, which they did in 1965 during studies of its molecular architecture. Reviewing Chinese work on insulin in 1974, the Nobel

Figure 40 The Strip-of-Sky Bridge on the Chengdu–Kunming Railroad in western China.

The subsidiary arches in the haunches of the main arch reflect the influence of the ancient Zhaozhou Bridge. The diesel locomotives used on this railroad, which opened in 1970, were all built in China.

(Illustration by Hazel Cotterell based on photographs in Mao Yi-sheng.)

prize-winner Dorothy Crowfoot Hodgkin attributed its success to the way Chinese scientists had 'integrated researches in different disciplines and laboratories'.[10]

Considering how traditional medicine and western medicine are used to complement one another, it may be significant that some of the most successful Chinese innovations in applications of microelectronics have included diagnostic instruments, ideas for which may well have been suggested by traditional practice. One is for diagnosing heart disorders, and another is a thermal sensing device used in detecting early-stage cancers, and in work on skin grafts.[11] Both these electronic devices won prizes at technology fairs in Europe in the 1980s.

The view that Chinese success with insulin synthesis was due to the integrated organization of the work brings us to a point concerning both China and Japan. Even when the technical content of work done in these countries appears very similar to its western counterpart, the institutional and organizational context may be very different, and this can direct innovation in new directions. In Japan, distinctive attitudes to industrial organization and management became very apparent when post-war recovery got under way in the 1950s. By the end of that decade, there was a carefully planned and very rapid expansion in a number of crucial industries, notably chemical, electrical and machine-tool manufacture. Steel, shipbuilding and motor-vehicles industries were also growing, heavily dependent on imported raw materials.

The enormous success of these industrial developments was quite clearly based on a distinctive style in the practice of technology, differing from the style of most western countries not only in organization, but also in ways of conceptualizing problems, attention to detail and attitudes to very fine work. The latter can be seen in the development of electric lamp manufacture in Japan as long ago as the 1890s. Light bulbs were soon being exported in large numbers, but significantly it was with bulbs of the smallest sizes (such as those for battery-operated torches) that Japanese industry was most successful. A possible explanation is that there was a transfer of skill from traditional production requiring fine, detailed work, including aspects of silk and porcelain manufacture. Similar factors may underlie the later growth of industries making miniaturized electronic components.[12]

Different questions arise about ways of conceptualizing problems and how this may be related to invention. A view which is frequently discussed is that Japanese creativity in basic research and invention has been very limited in comparison with the enormous national achievement in the practical application of new technology.[13] In the 1960s, for example, the Japanese electronics industry was responsible for many practical innovations based on the transistor, yet the invention of this device had occurred in America in 1948. It arose from work at the Bell Telephone Laboratories, related to the long series of researches into the structure of matter discussed earlier. Whilst the radio valves of 1904 and 1907 stemmed from the initial discovery of the electron, the transistor depended on later discoveries about the behaviour of electrons in solid material. The research at Bell was sponsored by the American military and was related to the use of 'solid-state' devices in radar equipment. The Bell management did not believe that transistors could ever work well enough to be used as a general replacement for conventional radio valves, and confined their attention to specialized applications, many for military computers. Thus in the late 1950s Americans made the first integrated circuits incorporating several transistors, but it was the Japanese who developed the first transistor radio for the ordinary consumer.

Similarly, the video cassette recorder was invented in America in 1956, but it was a Japanese firm which turned it into a consumer product, after twelve years' research, dramatically reducing its cost (by a factor of 100) and designing models suitable for use with domestic television sets.

These examples suggest that, if there is a difference between the creativity of research and development in Japan and that in the West, it may be that the Japanese are more practical and westerners are more inclined to pursue technical dreams and exotic inventions. One certainly cannot accuse either Japan or China of accepting the whole package of microelectronics ready-made from the West without active dialogue and innovation. The Japanese transistor radios and video recorders and the Chinese devices for medical diagnosis can all be understood as outcomes of dialogue of this sort. In China, the characteristic diagnostic methods used in traditional medicine probably contributed to a distinctive view of the potential of some kinds of electronic device. By contrast, the distinctive Japanese approach may have been partly due to social pressures which directed research away from the military context of many American inventions. During most of the 1950s and 1960s Japan did little or no military research. For some time after 1945, American regulation prevented it. Beyond that, many scientists felt revulsion toward this kind of research, largely because of experience during the Second World War. In the late 1960s the Japanese Physical Society asked defence scientists to refrain from contributing to its conferences. Moreover, throughout this period, and into the 1970s, the best engineering graduates in Japan went into production engineering and consumer manufacturing, while their contemporaries in America were being swallowed up by NASA and military research.

A deeper question concerning creativity in Asian science and technology has been raised by a Sri Lankan social scientist, Susantha Goonatilake.[14] He notes that in 1980 India had more trained scientists and engineers than Japan, and indeed more than most countries apart from America and Soviet Russia. But there were few major Indian achievements in science and technology to point to. Part of the problem was shortage of resources for research. A more basic difficulty, however, was that research workers in India felt that only by following recognized lines of research as established in the West would their work be 'legitimized' and gain recognition. This inhibited technological dialogue involving distinctively Indian points of view, with the result that creativity was 'aborted'.

If some aspects of Japanese technology earlier in the century were inhibited for similar reasons, the overwhelmingly impressive Japanese record in recent decades must surely have overcome this. China also has been more adventurous and successful in technological innovation than India, especially where the need for self-reliance has encouraged a dialogue with its own technological traditions and organizational genius. There has been dialogue of a similar kind in other cultures also, and it is the significance of this that forms the subject of the final chapter.

11 Survival technology in the twentieth century

The public health revolution

The culmination of the story told in previous chapters might seem to be the computer revolution of the 1970s and 1980s, practical uses of artificial satellites, and perhaps laser technology. It is here that some of the most striking technological dreams of the later twentieth century are working themselves out. But whilst computers, like engineering and the chemical industry in previous chapters, seem obvious topics for a history of technology, there have also been important developments in less spectacular technologies which are nonetheless essential to human survival, notably in relation to public health and agriculture. Moreover, developments in these technologies have made a bigger difference to people's lives throughout the world than most advances in electronics or engineering.

One example is provided by the substantial decline in deaths among young children now evident to a greater or lesser extent in most countries of the world. In Britain, the turning point came about 1900. Before that date, about 150 babies in every thousand died within a year of birth, and others died before reaching the age of five years. Thus a woman with three surviving children might have given birth to four or five. An infant mortality rate at about this level had persisted for several decades. Improvements in sanitation and water supply involving considerable engineering construction had led to a reduction in much adult illness, including typhoid and cholera, but deaths among infants remained high. When improvement began around the turn of the century, it came as the result of many factors, including better obstetric care, better hygiene, slowly improving housing conditions, better educated mothers and more adequate nutrition.

Parallel improvements occurred in other European countries and in North America, sometimes earlier than in Britain, and it became clear that the poor health of working people in the industrialized countries did not have to be regarded as inevitable. This fuelled campaigns for improved levels of welfare during the 1920s and 1930s. Then reconstruction in Europe after 1945 saw the rapid development of health services available to all. At the same time it appeared that worldwide campaigns against particular forms of ill health such as malaria and malnutrition were possible. The United Nations Food and Agriculture Organization was founded in 1945–6 to tackle food supplies and malnutrition. Its first director, John Boyd-Orr, had campaigned on the issue of malnutrition in British industrial cities in the 1930s – a period which had seen important discoveries about vitamins and protein, but which had also witnessed falling living standards due to unemployment. This was a formative experience for Boyd-Orr, strongly influencing his approach to the world situation.[1]

In the next three decades, there were reductions in infant mortality in many countries, most strikingly in China and East Asia, but with definite progress in India also. In Europe and North America, infant mortality rates fell to very low levels indeed. This should be regarded as one of the most significant technological achievements of all time. Where it was complemented by improved birth-control techniques, women were liberated from the burdens of frequent childbirth, and living standards rose because biological and economic resources were no longer needed to support children who died in infancy or later childhood. But it is a technological achievement which is hardly recognized as such because it is not associated with any clear symbol of progress or power. There is no single, striking innovation which contributed uniquely to this change. Many aspects of what happened were the work of women, so often undervalued, in their roles as mothers or as professionals in nursing, nutrition or related fields. In other words, a mixture of social, nutritional, domestic, medical and engineering developments had contributed, and while they were of the very essence of 'survival technology' (chapter 2), there was no invention nor any item of hardware to symbolize what had been achieved.

There was also a negative aspect of falling infant mortality. In some countries, conditions for the new generation were far from ideal. More children survived, but in a deteriorating environment. Pressures of 'modernization' led to the displacement of traditional birth-control practices before more modern methods could be introduced. Thus practices such as prolonged breast-feeding to inhibit further conception gave way as bottle-feeding was introduced. With the disturbance of other social customs, this meant many countries entered a period of sharp demographic transition with birth rates remaining high even though more children were surviving. Inevitably, populations increased fast, generating acute environmental stresses.

By the 1980s birth rates had declined substantially in East Asia, and were showing signs of falling in most other countries, except some in

Africa. Moreover, food production was increasing as fast or faster than population growth almost everywhere, again excepting Africa. The fact that the most populous countries, China and India, have been largely self-sufficient in food is another remarkable achievement of survival technology. Moreover, much of the necessary innovation has originated in those countries. Unlike innovations in electricity, engineering and aeronautics in the twentieth century, advances in agriculture and public health have not been wholly attributable to western technology.

A comparison between India and China demonstrates some aspects of this. At the end of the 1940s, both countries acquired new forms of government and greater freedom to shape policy according to national needs. But India was very open to western ideas whereas China was relatively isolated, especially after the sudden withdrawal of Russian advisers in 1960. Thus whilst India adopted western technologies in some areas of public health, such as malaria control, and in agriculture, with the 'green revolution', China was forced to depend much more on the innovative ability of its own people.

One striking example is associated with an innovation originating in the western chemical industry, the insecticide DDT, which was shown to be highly effective against malaria-carrying mosquitoes in southern Europe in the late 1940s. Soon schemes were underfoot to control malaria in India. A remarkable campaign mounted during the next two decades involved spraying the walls of every dwelling in India with DDT at regular intervals. This had an immediate impact on transmission of the disease, but nothing was done about the poor living conditions which made people vulnerable, nor were efforts made to encourage precautions that might have reduced mosquito breeding places near people's homes. Once mosquitoes acquired immunity to the insecticide, therefore, the disease could spread very easily again, and the number of cases increased substantially during the 1970s.

Meanwhile, China had to tackle malaria without being able to afford widespread use of DDT. The method was to hold numerous meetings in which people were informed of the environmental conditions under which this and other diseases spread. Many aspects of sanitation and drainage were examined, and teams were formed to fill in puddles or redesign water courses used for rice irrigation so that mosquito breeding places (and habitats for other pests) were greatly reduced. Housing very slowly improved and people were trained to take appropriate precautions. The impact on the spread of malaria was less dramatic than in India, but was probably more permanent, and the campaign was integrated with work on other public health issues connected with sanitation and hygiene. This stimulated the invention of several devices for treating human and animal excreta, including the Sichuan biogas systems discussed later in this chapter.[2]

Environmental technologies

Whilst malaria was prevalent in Africa also, the public health problem which most preoccupied Europeans working there during the colonial period was sleeping sickness (trypanosomiasis). The parasite which causes this is spread by the notorious tsetse fly, and is a scourge of cattle as well as humans. The conquest of the tsetse fly and the disease it carried became a major scientific goal for Europeans in Africa, who observed that sleeping sickness was steadily spreading into new areas, and who thought it the result of the 'backwardness' of African agriculture.

We now know that this assumption was almost the exact reverse of the truth. Before the colonial period, tsetse fly was well controlled in many parts of Africa. For example, in Botswana[3] herdsmen kept grazing lands free of the fly by burning the grass at certain seasons. This prevented the growth of shrubs and greatly restricted the habitat available to the insect. Comparable measures were used elsewhere, one particular operation being carried out in 1861 by people of Zulu descent living on the borders of Moçambique, who then successfully kept tsetse fly out of their area for over three decades.

There was a breakdown in these and other tsetse management arrangements in Africa following the colonial conquests of the 1880s and 1890s, and the arrival of a new cattle disease, rinderpest. This was accidentally introduced into Sudan by Europeans in 1889. It spread rapidly through all the cattle-raising areas of the continent, killing 90 per cent of livestock in some places. Many grazing lands became disused, enabling the bush to regrow, and thus provided an enlarged habitat for the fly.

Tackling the problem with false assumptions about why it was getting worse, colonial scientists 'almost completely overlooked the very considerable achievements of the indigenous people'. Admitting this, one of them sadly concluded that 'if we achieved anything at all, it was often to exacerbate the ills of the societies we imagined ourselves to be helping.'[4]

Not only were European assumptions about the local situation wrong, but so also was their belief in the universal applicability of all aspects of their science and technology. There is always a sense in which the fundamentals of science have universal validity. But in agriculture and public health, we require an approach which is also sensitive to local environments. The anthropologist Paul Richards talks about this in terms of 'ecological particularism',[5] implying that African success in the control of tsetse fly was due to empirical knowledge of the insect in specific local habitats, and was based on long experience of specific environments. This knowledge would be very 'particular' to those environments, and might be useless elsewhere. By contrast, European scientists working in Africa had much knowledge of general relevance, but insufficient that was locally specific.

In a wide range of other fields, including agriculture, soils, nutrition, mineral prospecting and herbal medicine, there are stories of western scientists undertaking research and only later discovering that relevant ecological information could have been learned from local Africans. For example, after a soil survey had been carried out in Ghana by American scientists at considerable expense, it was discovered that local farmers already recognized the main soil types the survey had identified, and could have saved the scientists much time if consulted earlier.[6] Obstacles to use of such knowledge are partly that it is not written down and can seem unsystematic, but also, like much 'folklore' about nature in Europe, it is sometimes expressed in terms of myths and proverbs.

To gain more perspective on the use of technology originating in other cultures, it is worth recalling from earlier chapters that at one time western civilization gained enormously through transfers of technology from Asia. However, this process had ended some time before 1900 because Europeans and Americans were entering new fields in electricity and chemistry which earlier Asian technology had never foreshadowed. Environmental technologies were quite a different matter, however, and in this area western equipment and methods were often specific to European or northeastern American environments. In major twentieth-century achievements which have required knowledge of other environments, whether arctic or tropical, westerners have sometimes borrowed heavily from people of other cultures with relevant experience. Thus when Robert Peary and Roald Amundsen made the first fully successful expeditions to the North and South Poles in 1909 and 1911 they used Inuit (Eskimo) technology, particularly for protective clothing (fur-lined anoraks) and for transport (dog-hauled sledges). Captain Scott certainly reached the South Pole without much use of these techniques, but his return journey ended in tragedy.

The green revolution

These points have a bearing on one of the major achievements of survival technology in the twentieth century, namely the 'green revolution'. At first sight, this is entirely the result of western innovation in what Andrew Pearse has called 'genetic-chemical technology'. The starting point was a series of developments in plant breeding and chemical fertilizer application located mainly in the United States. In the early 1930s, scientists working there produced a hybrid variety of maize (corn) which yielded much more grain per hectare than the varieties which farmers were currently growing. The hybrid caught on first in Iowa in 1934, and then more slowly spread to other states.

Repetition of this experience in other countries was masterminded by the Rockefeller and Ford Foundations. In 1943, four American plant geneticists were sent to Mexico by Rockefeller, and soon succeeded in producing improved varieties of wheat and maize to suit local conditions. Mexico's wheat output tripled in two decades. In 1962, the two foundations established the International Rice Research Institute in the Philippines. Within five years, a so-called 'miracle rice' had emerged. It produced greatly increased yields of grain provided it was grown with ample fertilizer and rigorous preventive measures against pests and diseases.

In India, the adoption of these new high-yielding cereals was hastened by an appalling drought during the years 1965–7. Grain production in the first drought year was 10 per cent less than in 1960–1, and when food aid was sought from the United States to make up the deficit, the aid was given on the understanding that India would introduce the new varieties of wheat, maize and rice on a large scale. This entailed a comprehensive programme for distributing chemical fertilizers and pesticides and for improving irrigation (often with the aid of electric pumps, as mentioned in the previous chapter). The new crops would probably have been tried anyway, but with greater caution and more careful investment in equipment and chemicals.

Grain output in India increased considerably over the next dozen years. Rice and wheat yields almost doubled, and since more land was sown with these crops (especially wheat), total output increased even more dramatically. India became self-sufficient in grain and in good years could even export some.

The conventional view is that this was an unambiguous technological triumph and that it was entirely due to the transfer of technology from American-inspired plant-breeding programmes. However, the point has been made in previous chapters that when new technologies are introduced into a country, their success almost always depends on the local innovations which they stimulate. These are usually of an *adaptive* nature, and help to match the transferred technology to local conditions. Occasionally, however, radically new ideas emerge.

Much of the adaptive work necessary to support India's green revolution was carried out by Indian plant breeders at local agricultural research stations. But these institutions were situated on good land, and it was experiment and innovation by ordinary farmers which ensured the success of the new crops on the poorer soils. Sometimes, indeed, the farmers chose to use crop varieties which had been rejected by official research. One rice variety developed by the plant breeders and then discarded was known as IR 24. Farmers who got hold of this liked it and multiplied the seed themselves, selecting plants which did well in local conditions. The result has been rice which is much more resistant to a local insect pest than are any of the varieties approved by the scientists.

Other local innovations included a method of weed control devised by farmers who could not afford to spray their crops with the recommended chemical herbicides.[7] This was a drastic but carefully timed harrowing operation. It depended on the observation that the most serious weed which had to be combated was especially vulnerable to mechanical damage a month after the crop was sown, but rice at that stage would not be harmed. Scientists were surprised when farmers explained this to them but confirmed its validity in their own experiments. Two of the scientists later commented that this was 'a turning point in making us realize that local practices are not altogether irrelevant'. Once again, local environmental knowledge had made an important contribution in awakening scientists to ecological particulars which their generalized approach might otherwise have missed.

The green revolution in India was mainly of benefit to farmers with large holdings. Many people with small plots were worse off. Many consumers who could not afford to buy wheat did not benefit at all, because there was no plant-breeding programme for the local coarse-grain crops on which they depended. It is instructive to observe, however, that output of many of the coarse grains has increased substantially. Two of the most important of these millet-type crops are jowar and bajra, for which output roughly doubled between 1950–1 and 1977–8. This was partly due to larger areas of land being cropped, and partly due to increased use of fertilizer and irrigation water. In addition, there was constant experimentation by farmers with the varieties grown. Here again, therefore, improvement in grain production was not solely the result of green-revolution technologies transferred from elsewhere. There was local development as well.

In China, improvements in output of cereals and pulses depended even more on local innovation, as well as on greater fertilizer use and extensions to the irrigated area, since China was relatively isolated during the early stages of the green revolution. Increased fertilizer applications were regarded as crucial, and many factories were built during the 1960s and 1970s to make nitrogenous fertilizer from synthetic ammonia. As already mentioned in chapter 10, some modern equipment was imported, but in addition, many small plants were built by local communes using equipment made in Shanghai. However, production of chemical fertilizer was never enough and it was always essential to make the fullest possible use of manure derived from human and animal excreta. There can be considerable health hazards in doing this, but they are much reduced by first 'digesting' the excreta in septic tanks. Many devices of this sort were invented in China, ranging from simple holding tanks located on the edges of rice fields to elaborate systems for treating effluent from pig farms. However, one of the most important innovations was a tank developed in Sichuan Province which was adapted for the collection of the methane gas generated during the digestion of the excreta. The gas was then used as a cooking fuel.

Techniques for making gas in this way had been the subject of experiment in China during the 1920s, and a company promoting them was founded in Shanghai in 1931 by Luo Guorui.[8] However, it was not until more than three decades later, with redesigned equipment, that 'biogas' came to be widely used. One factor then was that deforestation was a major problem in Sichuan, and it was policy to reduce the cutting of trees for fuel. Biogas production was relevant to a variety of environmental concerns in the area, therefore. By this time, biogas was being produced from animal excreta in Korea, Taiwan, Thailand and India as well as China. However, the apparatus in which the gas was generated in these countries was not as simple or as cheap to build as the equipment developed in Sichuan (figure 41), where many thousands of farming households were producing and using the gas by the end of the 1970s.

African agriculture

In Africa, just as in India, farmers did their own adaptive work on new cereal varieties introduced by western scientists. In Sierra Leone, for example, when a new rice variety was brought in, farmers tried it out on a small part of their land, and if it yielded well and its taste and keeping qualities were satisfactory, they added it to the repertoire of different strains of rice they regularly planted.[9]

The deliberate way in which farmers tested new varieties is shown by the context of a local word *sainu*, meaning 'experiment' or 'try out'. When farmers encountered an unfamiliar rice, they tried it in a systematic way, planting the seed on a carefully selected plot, and using a suitable container to measure the amount sown. After harvest, the same container measured the amount of grain produced, and output was compared with input.

One reason why there was no green revolution in Africa is that plant breeders did little work on locally important crops. For example, in the dry areas on the fringes of the Sahara Desert, millet and sorghum are

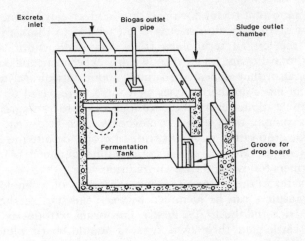

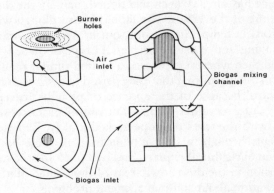

Figure 41   Concrete tank for 'digesting' sewage to produce biogas, developed in the Sichuan province of China, with (below) the earthenware cooking stove in which the gas is burned.

Pig excreta and human sewage is fed through the inlet on the left of the tank (top), with chopped straw and sufficient water to retain the fluid level above the various openings. The sludge extracted from the outlet is a relatively safe fertilizer for use on the fields.

(Illustration from McGarry and Stainforth, with acknowledgement to IDRC in Canada.)

the main cereals, and there is advantage in growing varieties which mature quickly. Then if the rains end early and drought sets in, there may still be a harvest. In one semi-desert region of northern Kenya in the early 1960s, women farmers introduced agronomists to quick-maturing sorghum strains which appear to have been previously unknown to science, and had certainly not been studied by the plant breeders.

A more important reason for the lack of a green revolution, however, was that western agriculturists too often tended to think in terms of *imposing* transfers of technology from supposedly more 'advanced' regions without allowing for local knowledge, experiment and innovation. Too often, also, they appeared to think about agricultural techniques developed in more northern climes as if these were based on universal principles to be applied everywhere without regard to 'ecological particulars'. For example, they tended to advocate large fields, each planted with a single crop rather than a mixture of crops, despite the ecological advantages of the latter and the good results obtained that way by African farmers following their own tradition.

Three phases in the attempt to impose transfers of technology under European auspices can be identified between the start of the colonial period in Africa and the 1970s. Firstly, Europeans introduced their own breeds of cattle and their own type of mould-board plough. One consequence has already been mentioned, namely the disastrous rinder-pest epidemic of the 1890s. The mould-board plough was well adapted to conditions in temperate Europe, but in Africa it was associated with very damaging soil erosion, first on European-owned farms in South Africa, and then wherever Africans acquired similar ploughs.

It was customary during this early period to dismiss African agriculture as 'backward' because of the absence of ploughs (except in Ethiopia), and also because so-called 'shifting cultivation' practices seemed primitive and fields with mixtures of crops looked untidy. Experience of tropical environments brought a slow modification of these opinions among a few colonial agricultural experts, and by the 1930s the salutary example of soil erosion in the dust-bowl areas of the United States had evoked a mood of caution about soil-management problems.

However, in the aftermath of the Second World War there was a new mood of confidence in the ability of modern science to solve such problems, and this led to more enthusiasm for transferring technology without much reference to local adaptation. The period from 1945 through to the 'Freedom from Hunger' campaign of the early 1960s saw a host of remarkable schemes for boosting food production in Africa. There were groundnut projects in East Africa, rice programmes in Nigeria and new irrigation structures. Mounting concern about malnutrition found expression in the idea that tractors could allow African farmers to grow more food for local consumption, and many were imported by aid agencies.

Nearly all these projects were disappointing and some were ludicrous failures. Tractors broke down, irrigation schemes were less productive than expected, there was a loss of diversity in cropping and a failure to appreciate the importance of trees in African agricultural ecosystems (partly because of the western habit of treating 'agriculture' and 'forestry' as separate disciplines). Even so, the idea that agricultural development should mean progress toward a European style of farming still has influence, and all too often this has led to 'the oversimplification of production systems and the "homogenization" of the landscape'. The most common result in eastern and southern Africa is 'the conversion of dry savanna woodland and complex pastoral and cropping systems into large tracts of monocropped maize interspersed with degraded grazing land'. One East African district where this has happened is noted equally for the severe poverty of its people and for 'the treeless, eroded landscape'.[10]

By the end of the 1960s, however, a third phase of thinking about agriculture was beginning. The argument was increasingly heard that the technology needed in Africa should be 'appropriate' in scale and expense when compared with the large and costly scale of earlier irrigation and tractor projects. E. F. Schumacher, author of *Small is Beautiful,* coined the term 'intermediate technology' after experience in Burma and India, and by the end of the 1960s his colleagues were into their first African project exemplifying this idea. Many other agencies involved in African development had already stopped thinking in terms of tractors, and were devoting more effort toward introducing seeds and fertilizers so that some of the new high-yielding crops could be grown.

All this was a step forward, but westerners were still thinking of arranging transfers of technology from outside Africa without asking what techniques Africans already possessed. Any 'relevant knowledge' of farmers was still felt to be 'delegitimized' by the assumed superiority of western science. However, the new emphasis on the appropriateness of transferred technology helped to focus attention on the actual circumstances of African farmers and on the ecological particulars of their local situations. This led to numerous discoveries, one being that many African farmers were women, and that some policies had failed – and are still failing – through being addressed solely to men.

Another discovery was that African agriculture included many ingenious techniques for water conservation and irrigation (figure 42). Some of these attracted attention because they involved very small-scale engineering such as 'stone lines' for water conservation in the Sahel region, or canals and terraced hillsides in Zimbabwe (reported by archaeologists) or simple bunds and sluices in rice-fields (as in the West African district discussed in chapter 6). Sometimes lever-and-bucket devices (shadoofs) were used in West Africa to raise water for irrigation. Often, however, irrigated agriculture in Africa has been inconspicuous, chiefly because so much of it is based on the exploitation of 'natural irrigation' on flood plains and valley bottoms and in swamps.

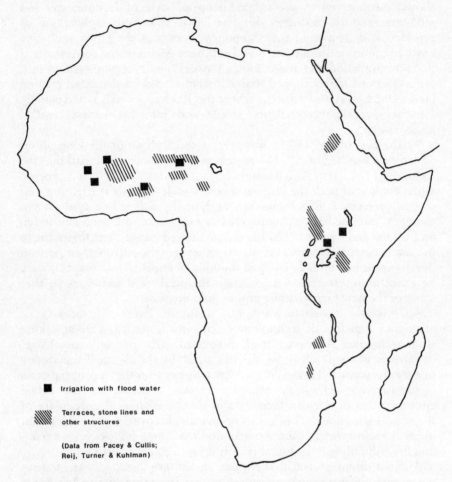

Figure 42 Distribution of selected water conservation and irrigation techniques in Africa south of the Sahara.

The techniques represented are appropriate mainly for the drier agro-ecological zones, and their distribution reflects the location of such zones. Techniques suitable for wetter areas, such as the rice systems mentioned in the text, are not represented.

European observers had overlooked these last aspects of African irrigation, partly as a result of the western tendency to equate 'technology' with 'engineering'. A different perspective is possible, however, if one redefines irrigation as 'any process, other than natural precipitation, which supplies water to crops'.[11] When one adopts this definition, there is no longer a sharp distinction between fields watered via canals or pumps, and lands where farmers exploit moisture supplied naturally, as on river flood-plains, or where there is runoff water from an adjacent hillside. In many places, there is a continuous spectrum from moist valley bottoms which can be treated as irrigated land, through seasonally moist fields on higher ground, to dry uplands, and the skill of the farmer does not lie in modifying the water supply to any part of the land but in choosing cropping strategies which can make best use of these varying conditions. The result can often be a more efficient use of land, water and labour resources than in many engineered irrigation schemes.

Other African farming methods which have recently been rediscovered are intercropping – that is, growing mixtures of complementary crops within a field – and the integration of trees with other crops in various forms of 'multistorey farming' (figure 43). Both these techniques contrast sharply with the European concept of growing a single crop in a field that is cleared of all other plants. The bad effects of this kind of monocropping have already been mentioned. They arise to a large extent because of the long period after ploughing and sowing when the soil is exposed with no vegetation cover. Such exposed soil is extremely vulnerable to leaching and erosion by tropical rainfall and there are many areas where the European approach has led to serious degradation of land. That applies in South Asia, Central America and especially Brazil as well as much of Africa.

By contrast, African techniques for growing mixtures of crops provide a denser coverage of the ground for a longer period in each year, because of the overlapping seasons of the crops. Thus the soil is better protected from the rain. In multistorey farming (mentioned in chapter 4 as practised by the Maya civilization in Central America, and having parallels in Indonesia mentioned in chapter 2), many trees are retained when forested land is cleared for cultivation, and extra fruit trees are often planted. Thus cereals, vegetables and root crops are grown in an environment where trees give additional protection against erosion, whilst their leaf litter helps maintain soil fertility. It is striking that farmers in three continents evolved this principle independently, but that it is only now being rediscovered by Europeans who previously thought that their cleared-field style of monocropping was a universally applicable technique. Such experience should clarify the point that whilst scientific knowledge of such things as photosynthesis and genetics has universal

Figure 43   Multistorey farming as practised in some parts of West Africa.
   The upper storeys are represented by oil-palm and banana crops. Below that
is the cereal crop, maize (corn). Nearer ground level are beans and melons,
whilst in the basement are the main root crops. Left to right, the latter are yams,
cocoyams and cassava. This somewhat compressed view does not reflect
realistic spacing, in which maize would be in open glades and more shade-
tolerant species under trees.
   (Illustration by Hazel Cotterell, based on discussions of multistorey farming by Richards,
also Niñez.)

validity, methods of planting crops and providing them with water and
nutrients must depend on the ecological particulars of local environments.

Innovation in Africa

One common opinion among Europeans during the colonial period was not only that African agriculture was 'backward', but also that local techniques never changed or developed. Taking a longer view, however, we find that far from being resistant to innovation, African agriculture has a long history of vigorously adopting new crops, including Southeast Asian crops such as the banana from about AD 500, and American crops such as maize and groundnuts from about 1500 (chapters 1 and 4).

Some aspects of more recent innovation can be appreciated from what has been observed in Kenya as farming families have responded to the gradual disappearance of wild foods and fruits due to the clearance of forests. In many cases, the threatened extinction of useful plants was prevented by domesticating them and then growing them in household gardens.[12] This entailed a series of trials and experiments which began when seeds or cuttings from a wild plant were first grown on a farmer's land. For this initial trial, the farmer would choose a site which was as similar as possible to sites where the wild plant was growing. Thus if the plant had been found near a stream (or under trees), the farmer's first attempts to grow it were also near a stream (or trees). But when the first crop had matured and produced seed, the farmer would experiment by planting the seed for another crop in an environment slightly more typical of the rest of her farm or garden. In collecting seed, she would also *select* it from plants which did particularly well in each environment. By experimental planting such as this, the farmer gradually produced a new variety of the plant, selected for compatibility with domestic cultivation. Several vegetables, root crops and fruit trees were domesticated in western Kenya during the 1960s and 1970s through processes such as this, and in response to pressures from deforestation.

Active innovation has also been a feature of other technologies practised in Africa. In metalworking, for instance, although the traditional techniques of bronze founders, coppersmiths and blacksmiths have all but disappeared in the face of cheap, imported metal goods, new activities have developed. One example discussed elsewhere[13] is the manufacture of sheet-metal water tanks in Kenya, which has become a significant local industry. A second and related example, again from Kenya, is production of the small cooking stoves used by 85 per cent of urban households, and by rural people also. Like the water tanks, the stoves are made of sheet steel, but this time usually scrap. They originated in a transfer of technology from India in the 1920s, when local Africans learned the method of manufacture from Indian businessmen living in Kenya. Techniques were adapted to materials and tools currently available, and an industry developed which in 1985 consisted of some 8,000 self-employed artisans.

The design of these stoves has become a critical aspect of local survival technology because deforestation has led to a shortage of charcoal and firewood, the fuels usually burned in them. People must either walk long distances to collect firewood, or pay more than they can easily afford to buy fuel. One response has been to design a more efficient version of the standard metal stove, so that a meal can be cooked with much less fuel. An important innovation which contributed to this aim was to make a ceramic lining for the stove capable of being produced by local potters (figure 44).

Although several organizations interested in 'appropriate' or 'intermediate' technology have played a part in developing the improved stove, it appears that a major source of ideas was a local man named Max Kinyanjui. He invented a modified stove in 1979 to meet cooking requirements in his own home. Realizing that the stove could be more widely useful, he initiated a project with six stove-makers in 1981. Methods of manufacture had to be adapted to the very basic tools and difficult conditions under which these men worked. Even so, many thousand stoves of the improved design had been made in Kenya five years later, and the innovation was thus of considerable economic significance locally.[14]

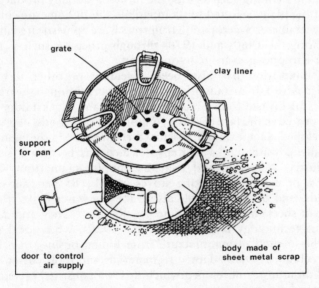

Figure 44 The new type of charcoal- or wood-burning cooking stove now widely used in Kenya, showing the clay ceramic lining.
(Illustration by kind permission of Intermediate Technology Publications)

If we analyse other twentieth-century innovations originating in Africa, we notice that whilst some are a response to economic pressures and to changing environmental conditions such as deforestation, many others reflect a stimulus effect from western technology, or some other form of interaction between indigenous and alien techniques. Sometimes the result is only of minor and local interest, as when a Kenyan woman responded to American scientists in a forestry programme by devising new ways of using a tree's leaves in compost, or when a Zulu herbalist incorporated a European type of spinach into the collection of plants used in his practice. But such episodes are of the very essence of 'technological dialogue', and sometimes it is western experts who have learned from observation of African techniques. Innovations which have resulted from these 'transfers' from Africans to westerners include improved grain stores,[15] water tanks made partly from local materials,[16] a basic technique for use with a West African soil and water conservation project,[17] and methods for control of an insect pest in Nigeria.[18]

However, in a particularly important book, the anthropologist Paul Richards indicates that the most potent stimulus to come from African technology into formal agricultural research is of quite a different order. A new appreciation of African techniques is beginning to shift the whole paradigm of tropical agriculture from a western model based on monocropping and 'soil and water engineering' to a completely different approach. With more attention paid to local techniques, and with active dialogue between innovating farmers and agricultural scientists, Richards sees scope here for an 'indigenous agricultural revolution' in Africa. Key elements in this would be integration of trees and field crops and irrigation with minimal engineering.[19]

The emergence of research in 'agro-forestry' is one sign that some western agricultural scientists are moving in this direction. Experiences in Indonesia and historical discoveries about Maya cropping in Central America have contributed to this, but Africa has been the scene of most activity, as well as being the locus of several innovations.

The best-known, if not necessarily the most useful, of such innovations is alley cropping, which originated with the observation that certain farmers in eastern Nigeria planted their fallow land with a quick-growing tree species (*Acioa barteri*) in order to speed up the regeneration of the soil.[20] In the 1970s, this practice was studied by B. T. Kang, an Indonesian scientist working at an international research institute in Ibadan. He saw that one did not have to plant just trees when land was to be fallowed. They could also be planted *with* a crop such as maize, in alternate rows or alleys. The trees would continue to grow after the maize had been harvested, protecting the soil from erosion and maintaining fertility with nutrients provided by their leaf litter. In the next season, when a new cereal crop was to be planted, the trees could be pruned back to give the crop ample light, the cut leaves and twigs being used for fuel and fodder. The trees would regrow, but mainly after the crop had been harvested.[21]

Conclusion: technology in world civilization

The history of technology, as presented in earlier chapters, offers many examples of innovation resulting from interaction between different kinds of knowledge and technique. Often, the 'transfer' of technical knowledge or equipment from one country to another has initiated a process of modification and adaptation from which new and inventive ideas have emerged. This process can be characterized as a 'dialogue', but how it proceeds must clearly depend on the knowledge and skill of the people who are encountering the transferred technology for the first time. When these people are already developing related techniques, their dialogue with the new technology is likely to be especially creative. If they have no relevant experience, but are nonetheless interested, they may still respond in the manner of a dialogue even if largely in non-technical ways, perhaps related to social arrangements for use of equipment. Either way, the experience, skill and inventive imagination which people contribute from their own cultural background is crucial.

For example, in Europe during the twelfth and thirteenth centuries, the transfer of technology from Islamic countries and China was particularly stimulating because Europeans already had a comparable culture of mechanical innovation which was itself evolving rapidly (chapter 3).

In India during the eighteenth century, shipbuilding techniques transferred from Europe were used creatively by local craftworkers in some of the biggest and best ships in the world, largely because India had a long shipbuilding tradition of its own in which there had been active innovation since at least the fifteenth century[22] (chapter 7).

In Japan, the adoption and development of technologies transferred from the West may seem to have begun abruptly soon after the Meiji restoration of 1868. However, the reality is that Japan was able to use those technologies effectively because a technically innovative culture had been evolving for a long time. This had been stimulated by a selective adoption of western techniques since the sixteenth century, and by a much longer tradition of borrowing and developing technology from Korea and China (chapter 5).

Similarly, the modern industrial success of Korea, Taiwan and other 'Pacific-rim' countries is related to the fact that Korea, in particular, has distinctive technological traditions of its own,[23] and other countries in the region are heirs to China's long technical experience.

In China itself, whilst the creative use of modern technology may appear to have begun only after 1949 under the government of Mao Zedong, successful developments in many branches of engineering, industry and medicine have been strongly influenced by China's own background in these areas of skill, and are not simply the result of 'transferred' techniques. During the nineteenth century, ultra-conservative government policies and a long period of civil war may seem to have nearly extinguished the nation's technological tradition. But there were some Chinese industrial initiatives (chapter 9), and from about 1900 engineering workshops associated with shipping companies, railroads and textile factories became significant for the revival of Chinese mechanical skills, and for coming to terms with 'transfers of technology' from the West. Workshops in the Shanghai area proved to be especially important in building up technological expertise. By 1933 there were 173 workshops there of a size to employ more than 30 people. One was the Talung Machinery Works, founded in 1902 simply to repair imported machines, but by the 1930s manufacturing spinning machinery. During the 1950s and 1960s, technicians from this and other Shanghai factories made a crucial contribution to the development of new engineering industries.[24]

Africa has been an instructive case-study for this final chapter in being a continent which, quite unlike China and Japan, has no long history of mechanical invention, and no heritage of nineteenth-century industrial workshops. Gold mining, metalworking, bronze-casting, sculpture and wire bracelet-making were part of some cultures, and have conditioned responses to modern technology – as when boys make wire models of automobiles, and young men from cultures with a tradition of fine sculpture prove to be good at three-dimensional thinking in the context of motor repairs and mechanical engineering. But these are small matters. The greatest technical expertise of many African cultures has been the survival technology they have developed to cope with their especially demanding environments. Recognition of the possibility of a dialogue between this fund of expertise and western agricultural science has come very late, after tremendous ecological damage has been done by inappropriate techniques, unbalanced development and population growth, but still offers more hope for the future than any pretence about the transfer of unmodified western technologies.

The tragedy for large parts of Asia is that just as the process of dialogue with the West had begun to bear industrial fruit, serious problems with regard to survival technologies began to appear. China's burst of industrial development in the 1980s was accompanied by air pollution, deforestation and worsening soil erosion, which have begun to threaten food production. Yet China has its own expertise in survival technologies which places a high value on trees, on recycling plant nutrients in manures and on deep insights in nutrition. Until the early twentieth century, knowledge of nutrition in China was far more discriminating and relevant to the maintenance of health than the crude ideas which prevailed in the West. Only with discoveries which led to the recognition of vitamins in 1911, and research on their significance in the 1920s and 1930s, could western scientists begin to offer equally informed dietetic advice. And some of that research was done in India, learning from Indian food practices.[25]

All these episodes in the long history of technological dialogue played their part in the development of what is increasingly becoming a single world civilization, albeit embracing many cultures. It is a world civilization partly in terms of the scope of economic and military power, and of trade and communications. Its significance is also apparent when we observe how many artefacts such as automobiles and computers are in worldwide use, with few design differences adapted to contrasting regions. For some commentators, this coming-together of civilizations means that distinctive approaches to technology in different parts of the world are no longer relevant. They argue that modern technology is based on universally applicable principles. For example, Joseph Needham, the historian who has so strongly emphasized China's contribution to early science and technology, believes that modern Chinese science is no longer distinctive because it has merged with 'universal world science'.[26]

A valuable corrective to that outlook comes from an Indian commentator who has spoken of 'the necessity of rejecting the western pretence of universalism'. Non-western cultures should develop 'pluralism' in their approach to technology, drawing on their own knowledge and skills as well as the western kind. By so doing, the non-western cultures may 'help westerners themselves in dealing with the new crop of problems they now encounter'.[27]

There is, of course, a vast body of scientific knowledge which ought to be accepted as universal (though always subject to revision). However, it is also evident that in many complex aspects of medicine, environmental technology and the application of electronics or engineering, the sciences that are 'universal' are not fully comprehensive. They do not cover all the particulars of specific situations. They do not allow for the ways in which the *practice* of technology involves social choices and managerial procedures, and is thereby culturally conditioned. It would be folly to deny the value of the universal elements in science and technology, but it is equally important to seek a balance between the universal and the particular.

For these and other reasons, one should not expect Chinese traditional medicine to merge with western medicine as universal science develops. One should not expect microelectronics or civil engineering to be used in China in exactly the same way as in the West, and indeed the traveller in China is struck by a local style in modern engineering (for example, using very refined arched bridges where western engineers would use box girders). Differences such as this are one reason why technological dialogue continues to be important, and why transfers of technology need to be complemented by local adaptation *and* invention. Other, more vital reasons are evident in situations where sensitivity to the ecological and social particulars of local environments is critical, especially in agriculture and public health technology.

There are, of course, many factors which have tended to obstruct technological dialogue or make it less fruitful. In China, the repression of civil liberties is having that effect. In India, deindustrialization eroded skills and self-confidence (chapter 7). A similar erosion of cultures, economies and skills has continued in many countries during the twentieth century. Africa may have the potential to generate its own agricultural revolution, but economic pressures favouring certain cash crops, and multinational corporations pushing their own agro-industrial packages, militate against it. New techniques in agro-forestry and multistorey farming may open the door to ecologically sensitive development in the rain-forest areas of Amazonia as well as Africa, but the powerful economic interests which prefer to destroy the Brazilian forest may continue to prevail.

However, it remains true that a 'powerful source of development' is 'the exchange of all things original that are created independently by each nation'.[28] With regard to technology, we have seen evidence of this over a long period of history. However, it has also been clear from these last pages that it is not only an exchange or dialogue between nations that has been important. There is also a dialogue within each society between people with different kinds of knowledge and experience, women as well as men, artisans (and farmers) as well as scientists. Related to this is a dialectic within science and technology between the universal and the particular; between survival technology and spectacular, symbol-creating developments; and between dreams and practical needs. In the past, much innovation has been stimulated by interactions of these kinds between and within cultures. The dialogue continues. It is of particular relevance for environmental and human aspects of technology, but is a source of creativity in high technology also.

Notes

NB The bibliography (p. 218) gives details of publishers and titles of articles where they are not quoted here, and also provides a full reference for authors quoted by name only in the comments on background material.

1 An age of Asian technology, AD 700–1100

Background information on **China** used in this chapter has been drawn from works by Bray, Elvin, Hua Jueming, also Rostoker and co-authors. On the **Islamic world**, the most useful works consulted were by al-Hassan and Hill, Norman Smith, Watson, also Wulff.

1 W. H. McNeill, *The Pursuit of Power*, 1983, p. 25.
2 R. Hartwell, *Journal of Economic History*, 26 (1966), pp. 29–58.
3 J Needham, *Science and Civilization in China*, volume IV (part 3), 1971, pp. 320, 350–2.
4 D. Hill, *A History of Engineering in Classical and Medieval Times*, 1984.
5 The most generally useful book on this topic is C. G. F. Simkin, *The Traditional Trade of Asia*, Oxford: Oxford University Press, 1968.
6 Needham, *Science and Civilization in China*, volume IV (part 3), 1971, pp. 154, 197; also Tohru Ishino, *How the Great Image of Buddha at Nara was Constructed* (in Japanese). Tokyo: Komine Shoten Publishing Co., 1983.
7 Tsien Tsuen-Hsuin, *Paper and Printing*, in Needham (ed.), *Science and Civilization in China*, volume V (part 1), 1985, p. 149.

2 Before the Mongols

Background material on **India** and on the **Mongols** has come from many sources including Ikram, Mathur, Morgan, also Simkin.

1 I. Habib, 'Technological changes and society', in D. Chattopadhyaya (ed.), *Studies in the History of Science in India*, volume II, 1982, pp. 816–44. On the origin of the stirrup, see L. White, *Medieval Technology and Social Change*, 1962.
2 Habib, 'Technological changes and society', takes an opposite view to W. Born, in *CIBA Review* (Basle), no. 28, December 1939.
3 W. Willetts, *Chinese Art*, volume I, 1958, pp. 107–243.

4 F. Bray, *Agriculture*, in Needham (ed.), *Science and Civilization in China*, volume VI (part 2), 1984, pp. 134, 501 etc. For other aspects of 'survival technology', I am indebted to Anil Gupta (private communication).

5 N. J. Dawood (trans.), *The Thousand and One Nights*, Harmondsworth: Penguin, 1954.

6 P. Avery and J. Heath-Stubbs (trans.), *The Ruba'iyat of Omar Khayyam*, Harmondsworth: Penguin, 1981, quatrains 24, 29, 33–4, 105 etc.

7 A. Y. al-Hassan and D. R. Hill, *Islamic Technology: an illustrated history*, 1986, comment on p. 63.

8 Needham, *Science and Civilization in China*, volume IV (part 3), 1971, p. 32; also L. C. Goodrich, 'Cotton in China', *Isis*, 34 (1942–3), pp. 408–10.

3 Movements West, 1150–1490

On **Africa**, some general information has been drawn from Ajayi and Crowder, also Oliver and Fage. For **Asia** and **Europe**, much background on military invasion and the diffusion of technology has been filled in from Carman, Cipolla's *Guns and Sails*, Landes, Morgan, also Simkin.

1 Quoted by A. M. Watson, *Agricultural Innovation in the Early Islamic World*, Cambridge: Cambridge University Press, 1983, p. 102.

2 H. L. Roth, *Studies in Primitive Looms*, 1918, reprinted Bedford (England), Ruth Beam, 1977, p. 63.

3 Al-Hassan and Hill, *Islamic Technology: an illustrated history*, 1986, p. 192.

4 Ibid., p. 62.

5 See especially J. Gimpel, *The Medieval Machine: the industrial revolution of the Middle Ages*, 1977.

6 A. Pacey, *The Maze of Ingenuity*, 1976, chapters 1 and 2.

7 D. S. Landes, *The Unbound Prometheus*, 1969, pp. 29–30.

8 Needham, *Science and Civilization in China*, volume V (part 7), 1986, pp. 39–42.

9 This statement is querying the interpretation suggested by al-Hassan and Hill, *Islamic Technology*, 1986, pp. 113–14.

10 Ibid.; also M. Elvin, *Pattern of the Chinese Past*, 1973, p. 89.

11 Needham, *Science and Civilization in China*, volume V (part 7), 1986. See pp. 229, 293, 329 for the fire-lance gun, the 1288 gun, and the bottle-shaped gun respectively.

12 Ibid., pp. 575, 577.

13 Simkin, *The Traditional Trade of Asia*, 1968, p. 140.

14 Ibid., p. 137.

15 Al-Hassan and Hill, *Islamic Technology*, 1986, p. 153.

16 Needham, *Science and Civilization in China*, volume V (part 7), 1986, p. 570.

17 Comments on the 'psychological effect' of the noises made by cannon can be found in C. M. Cipolla, *European Culture and Overseas Expansion*, 1970, p. 36; also in Elvin, *Pattern of the Chinese Past*, and especially in R. M.

Schafer, *The Tuning of the World*, Philadelphia: Pennsylvania University Press, 1980.

18 Needham, *Science and Civilization in China*, volume IV (part 3), 1971, pp. 297, 491–2, 526.

19 Different views are presented by Sang-woon Jeon, *Science and Technology in Korea*, 1974, and by Chinese historians such as Liu Guojun and Zheng Rusi, *The Story of Chinese Books*, 1985.

4 Conquest in the Americas, and Asian trade

Background material on the **Americas** has been drawn from a miscellany of specialist sources including Boserup, Bradfield, Bühler, Ewell and Merrill-Sands, P. D. Harrison and Turner, also Marcus. See also the brief but perceptive overview in Gille's general history of technology.

1 M. D. Coe, 'Chinampas of Mexico', *Scientific American*, 211 (July 1964), pp. 90–8.

2 G. R. Willey, 'Introduction', in K. V. Flannery (ed.), *Maya Subsistence*, New York: Academic Press, 1982, pp. 4–6.

3 Norman Smith, *A History of Dams*, 1971.

4 Needham, *Science and Civilization in China*, volume IV (part 3), 1971, pp. 542–4.

5 W. H. McNeill, *Plagues and People*, 1977, pp. 203–4, 206.

6 J. M. Roberts, *The Pelican History of the World*, Harmondsworth: Penguin (1980), revised edn. 1983, p. 466.

7 Sources for the maritime history in this section include Needham, *Science and Civilization in China*, volume IV (part 3), 1971, pp. 379–699; R. Mookerji, *Indian Shipping*, 1912; also Simkin, *The Traditional Trade of Asia*, 1968.

8 G. A. Horridge, *The Lashed-lug Boat of the Eastern Archipelagoes*, 1982.

9 A. J. Qaisar, *The Indian Response to European Technology and Culture, 1498–1707*, 1982, pp. 22–3.

10 N. Smith, *A History of Dams*, 1971.

11 For mining in West Africa, see E. W. Bovill, *The Gold Trade of the Moors*, 1968. For the Zimbabwe mines, see R. Summers, *Ancient Mining in Rhodesia and Adjacent Areas*, 1969.

12 See estimates quoted by Bovill, *The Gold Trade of the Moors*, 1968, pp. 106–18, also by J. F. A. Ajayi and M. Crowder (eds), *History of West Africa*, volume I, 1971, p. 388.

5 Gunpowder empires, 1450–1650

Useful if brief background material for the whole of this chapter can be found in Cipolla's *Guns and Sails*. Very little is available in English on technology in the Ottoman Empire, but Atil is useful for context and comparisons with the applied arts.

1 McNeill, *The Pursuit of Power*, 1983, pp. 83–7.
2 Al-Hassan and Hill, *Islamic Technology*, 1986, pp. 109, 117–18.
3 Marshall G. S. Hodgson introduced the term 'gunpowder empire' as the title of volume III in his *The Venture of Islam*, 3 volumes, Chicago: Chicago University Press, 1975.
4 McNeill, *The Pursuit of Power*, 1983, p. 148; also McNeill's *Plagues and People*, 1977, pp. 232–3.
5 On *wootz* steel, see J. Needham, *The Development of Iron and Steel Technology in China*, 1958, pp. 14–15; also Dharampal, *Indian Science and Technology in the Eighteenth Century*, 1977, pp. L–LII.
6 B. Bronson, *Historical Metallurgy*, 21 (1), 1987, pp. 8–15.
7 A. North, *An Introduction to Islamic Arms*, 1985, p. 9.
8 C. S. Smith, *A History of Metallography*, 1960, pp. 23–6.
9 T. Raychaudhuri and I. Habib (eds), *Cambridge Economic History of India*, volume I, 1982, pp. 313–14; also E. Atil, *The Age of Sultan Suleyman the Magnificent*, 1987, pp. 24, 26.
10 On dyeing, see M. Gittinger, *Master Dyers to the World*, 1982; J. Irwin and K. Brett, *Origins of Chintz*, 1970; G. Schaefer, *CIBA Review* (Basle), no. 39, May 1941.
11 Wulff, *The Traditional Crafts of Persia*, 1966, p. 282.
12 I owe a personal debt to the late W. V. Farrar for an introduction to the history of technology in Japan. Background information for these paragraphs has also been drawn from Bray (both titles listed in the bibliography), Chibbett, Cipolla's *European Culture and Overseas Expansion* (1970), Jeon, also Sansom.
13 K. Kubata, *Tatara Process Iron-making in Japan* (in Japanese), Tokyo: Komine Shoten Publishing Co., 1982.

6 Concepts in technology, 1550–1750

A discussion parallel with this chapter, but giving more detail of the European background, is to be found in A. Pacey, *The Maze of Ingenuity*, 1976, chapters 3 and 4.
1 The most comprehensive work on printing in Asia is Tsien Tsuen-Hsuin, *Paper and Printing*, in Needham (ed.), *Science and Civilization in China*, volume V (part 1), 1985. See also D. Chibbett, *The History of Japanese Printing and Book Illustration*, 1977.
2 Tsien Tsuen-Hsuin, *Paper and Printing*, 1985; also Simkin, *The Traditional Trade of Asia*, 1968.
3 Needham, *Science and Civilization in China*, volume II, 1956, p. 163; also H. Redner, 'The institutionalization of science', *Social Epistemology* (London), volume I, no. 1 (1987), pp. 37–59.
4 B. Gille, *The History of Techniques*, volume I, 1986, pp. 380–407.
5 C. A. Alvares, *Homo Faber: technology and culture in India, China and the West*, 1980, pp. 95–6.
6 Schaefer, *CIBA Review* (Basle), no. 39, May 1941.

7 Needham, *Science and Civilization in China*, volume I (1954), p. 146.
8 Cipolla, *European Culture and Overseas Expansion*, 1970, p. 159.
9 Al-Hassan and Hill, *Islamic Technology: an illustrated history*, 1986, pp. 11, 16, 69.
10 C. Merchant, *The Death of Nature: women, ecology and the scientific revolution*, London, Wildwood House, 1982, p. 220.
11 McNeill, *The Pursuit of Power*, 1983, pp. 127–39.
12 W. Schivelbusch, *The Railway Journey: the industrialization of space and time*, New York and Leamington Spa: Berg, 1986, pp. 150–8.
13 P. Worsley, *The Three Worlds: culture and world development*, 1984.
14 D. C. Littlefield, *Rice and Slaves*, 1981.
15 W. H. Chaloner, *History Today*, 3 (1953), pp. 778–85.
16 S. Bhattacharya and B. Chaudhuri, 'The regional economy . . . Eastern India', in D. Kumar and M. Desai (eds), *Cambridge Economic History of India*, volume II, 1983, p. 286.
17 Chaloner, *History Today*, 3 (1953).
18 A. P. Wadsworth and J. de L. Mann, *The Cotton Trade and Industrial Lancashire*, 1931, p. 433. For more detail of the Paul and Wyatt machine, see R. L. Hills, *Power in the Industrial Revolution*, 1970.
19 R. S. Fitton and A. P. Wadsworth, *The Strutts and the Arkwrights*, Manchester: Manchester University Press, 1958, pp. 64, 226.

7 Three industrial movements, 1700–1815

Background reading on the industrial revolution in Britain and its ramifications in India has included books by Hobsbawm, also Landes on **Britain**, and Braudel, Chaudhuri, also Simkin on **India**. See also Kumar and Desai, and works on specific Indian industries by Gittinger (dyeing), Irwin and Brett (calico painting), Ray (alum production). and Mookerji, Qaisar, also Wadia (shipbuilding).

1 Boserup, *Population and Technology*, 1981, pp. 89–90, 112–17.
2 Elvin, *The Pattern of the Chinese Past*, 1973, pp. 307–8; also Bray, *Agriculture*, in Needham (ed.), *Science and Civilization in China*, volume VI (part 2), 1984, pp. 507–8.
3 Elvin, *The Pattern of the Chinese Past*, 1973, pp. 285–6; also Wagner, *Historical Metallurgy*, 18 (1984), pp. 95–104.
4 An important source of background material for this section is C. K. Hyde, *Technological Change and the British Iron Industry*, 1977.
5 Pacey, *The Maze of Ingenuity*, 1976, chapter 4.
6 Boserup, *Population and Technology*, 1981, pp. 107–8.
7 B. Trinder, *The Industrial Revolution in Shropshire*, 1973, p. 360.
8 A. Juvet-Michel, *CIBA Review* (Basle), no. 31, March 1940.
9 Quoted by A. E. Musson and E. Robinson, *Science and Technology in the Industrial Revolution*, Manchester: Manchester University Press, 1969, p. 343n.
10 Ibid., p. 252.

11 F. Braudel, *Civilization and Capitalism: Volume 3, The Perspective of the World*, 1984, p. 522. India's role is also clearly recognized by Wadsworth and Mann, *The Cotton Trade and Industrial Lancashire*, 1931.

12 This is the implication of Braudel's views (ibid.), and the relevant data are summarized by D. Kumar and M. Desai (eds), *Cambridge Economic History of India, Volume II, 1757–1970*, 1983.

13 Ikram, *Muslim Civilization in India*, 1964, p. 274.

14 S. Goonatilake, *Aborted Discovery*, 1984, p. 93; also Hobsbawm, *Industry and Empire*, 1968, p. 49, both use the term 'deindustrialization'.

15 Herman Melville, *Redburn*, London and New York, 1849 (and many modern editions), chapter 34. The point about coir rope is confirmed by Qaisar, *The Indian Response to European Technology and Culture, 1498–1707*, 1982, pp. 28–30. Although it is a novel, Melville's book is based on experience during his own journey to Britain in 1839.

16 Quoted by Simkin, *The Traditional Trade of Asia*, 1968, p. 287.

8 Guns and rails: Asia, Britain and America

Background material on **India** in this chapter has been drawn from Dharampal, Egerton, Pant, Sarkar, also Weller. On ploughs and winnowing machines in China, India and Indonesia, it has been instructive to compare Bray's *Agriculture* with Dharampal, also Thorburn. Background for the **United States** has been provided by Hopkins, Hunter, also Morison (on specialized topics) and by the two Rosenberg books listed (for overall perspectives). In addition, I am indebted to experience of teaching Open University course A281, 'Technology and Change, 1750–1914'.

1 This is evident from illustrations given by G. N. Pant, *Studies in Indian Weapons and Warfare*, 1970, and from the efforts of Wellesley to improve gun carriages manufactured in India quoted by J. Weller, *Wellington in India*, 1972.

2 Quoted by R. G. S. Cooper, 'Arthur Wellesley's encounters with Maratha battlefield sophistication during the Second Anglo-Maratha War', paper read at the First Wellington Congress, 1987. (I am also indebted to Randolf Cooper for private communications on this matter.)

3 Needham, *Science and Civilization in China*, volume V (part 2), 1974, pp. 212, 220.

4 Ibid., volume VI (part 2), 1984, pp. 375–7.

5 Ibid., volume IV (part 3), 1971, p. 149.

6 James Anderson *Report on a Bridge of Chains to be thrown over the Firth of Forth at Queensferry*, 1818. The relevant Telford notebook is held by the Institution of Civil Engineers in London and is quoted by D. B. Hague, *Conway Suspension Bridge*, London: National Trust, no date, c.1970. See also H. Douglas, *Crossing the Forth*, London: Robert Hale, 1964.

7 L. T. C. Rolt, *George and Robert Stephenson*, London: Longman, 1969.

8 R. Fremdling, 'Railroads and German economic growth', *Journal of Economic History*, 37 (1977), pp. 583–601.

9 J. Nadal, 'The failure of the industrial revolution in Spain', in Carlo M. Cipolla (ed.), *The Fontana Economic History of Europe*, volume IV, part 2, London: Fontana/Collins, 1973.

10 This section draws heavily on the important book by D. R. Headrick, *The Tools of Empire: technology and European imperialism in the nineteenth century*, 1981.

11 R. A. Wadia, *The Bombay Dockyard and the Wadia Master Builders*, 1957, p. 330.

12 Pant, *Studies in Indian Weapons*, 1970, whose comments are put into perspective by discussion of the same theme by Headrick, *The Tools of Empire*, 1981, p. 90.

13 Headrick, *The Tools of Empire*, 1981, pp. 99–100; N. Rosenberg, *Technology and American Economic Growth*, 1972: also D. A. Hounshell, *From American System to Mass Production*, Baltimore: Johns Hopkins University Press, 1984.

14 For this section, see Kumar and Desai (eds), *Cambridge Economic History of India, Volume II, 1757–1970*, 1983; and for a specific case study, F. R. Harris, *Jamsetji Nusserwanji Tata: a chronicle of his life*, Bombay: Blackie & Son, 1958.

9 Railroad empires, 1850–1940

Background on **Japan** important for this chapter is based on standard economic histories by Lockwood, also Macpherson, plus Chamberlain's work on Manchuria, and on studies of 'choice of technique' by Fei and Ranis, Okita, also Ranis. Studies of European engineers working in Japan have also been consulted, including Bates, also Jones.

1 The main source for this section is J. N. Westwood, *A History of Russian Railways*, 1964.

2 S. Okita, 'Choice of technique – Japan's experience', in Kenneth Berrill (ed.), *Economic Development with special reference to East Asia*, 1966; W. W. Lockwood, *The Economic Development of Japan*, expanded edn. 1968.

3 J. C. H. Fei and G. Ranis, *American Economic Review*, 54 (1964), pp. 1,063–8.

4 Chamberlain, *Japan over Asia*, 1938.

5 On China's early industrial development, see A. Feuerwerker, *China's Early Industrialization*, 1958.

6 On the symbolism of railroads, see especially Leo Marx, *The Machine in the Garden*, 1964; also Headrick, *The Tools of Empire*, 1981 (notably pp. 183, 187, 194). For biographies of engineers who developed the symbolic aspect, see J. Harriss, *The Eiffel Tower: Symbol of an Age*, 1976; also L. T. C. Rolt, *Isambard Kingdom Brunel*, London: Longman, 1957.

7 Paul Theroux, *The Great Railway Bazaar*, London: Hamish Hamilton, 1975, chapter 17.

8 Marx quoting Walt Whitman in *The Machine in the Garden*, 1964, pp. 223–4.

9 J. Thomson, *China, the Land and its People*, 1873, reprinted 1977, p. 98 (plate 90).
10 Natesa Sastu, 'The decline of South Indian arts', *Journal of Indian Art*, 3 (no. 28), 1890, p. 23.
11 H. Carter, *Wolvercote Mill: a study of paper making at Oxford*, Oxford: Oxford University Press, 1957.
12 Trinder, *The Industrial Revolution in Shropshire*, 1973, p. 176 (2nd edn., p. 104).
13 Marx, *The Machine in the Garden*, 1964, p. 170.

10 Scientific revolutions and technical dreams

Background material on the history of the chemical and electrical industries has come from books by Dunsheath, Landes, Morison, Rosenberg (both titles listed), Scott, Sykes, also Williams. On research laboratories and technical education, information has been drawn from Cardwell, Farrar and Pacey, also Langrish.

1 Simkin, *The Traditional Trade of Asia*, 1968, p. 291.
2 F. Klemm, *A History of Western Technology*, 1959, pp. 342–7.
3 Harriss, *The Eiffel Tower: symbol of an age*, 1976.
4 F. I. Ordway, and M. R. Sharpe, *The Rocket Team*, London: Heinemann, 1979, p. 361.
5 For non-specialist background material for this section, I have used: L. F. Haber, *The Chemical Industry during the Nineteenth Century*, Oxford: Clarendon Press, 1958; also S. Katz, *Classic Plastics from Bakelite to High-Tech*, London: Thames and Hudson, 1984; and for radiochemistry, R. Jungk, *Brighter than a Thousand Suns: a personal history of the atomic scientists*, London: Gollancz and Hart-Davis, 1958.
6 P. Pringle and J. Spigelman, *The Nuclear Barons*, 1981, pp. 136, 137.
7 A. Rahman, 'Review of *Homo Faber*, by Claude Alvares', *Technology and Culture*, 23 (1982), pp. 479–81. Other quotations here are from Pringle and Spigelman, *The Nuclear Barons*, 1981, pp. 174–7, 391–9; P. D. Henderson, *India: the Energy Sector*, Delhi: Oxford University Press for the World Bank, 1975; also O. Marwah, 'India's nuclear program', in W. H. Overholt (ed.), *Asia's Nuclear Future*, Boulder (Colorado): Westview Press, 1977.
8 Wagner, *Historical Metallurgy*, 18 (1984), pp. 95–104.
9 Mao Yi-sheng, *Bridges in China, Old and New*, Peking (Beijing): Foreign Language Press, 1978, p. 32.
10 D. Crowfoot Hodgkin, 'Chinese work on insulin', *Nature* (London), 255 (1975), p. 103. See also A. Kleinman, P. Kunstadter, E. R. Alexander and J. L. Gale (eds), *Medicine in Chinese Cultures*, 1975.
11 Zhang Huafeng, 'Beijing's High-tech corridor', *China Pictorial*, No. 9 (Sept.) 1988, pp. 38–41.
12 On transistors, see F. Malerba, *The Semiconductor Business*, London: Francis Pinter, 1985. On light bulbs, see Lockwood, *The Economic Development of Japan*, 1955; expanded edn. 1968.

13 P. M. Boffey, 'Japan on the threshold of an age of Big Science', *Science*, 167 (1970), pp. 31–5, 147–52, 264–7.
14 Goonatilake, *Aborted Discovery*, 1984, p. 104.

11   Survival technology in the twentieth century

Background information on the 'public health revolution' comes from many sources, of which three are Gordon Harrison (on malaria), Thomas McKeown, also McNeill (*Plagues and People*). On the 'green revolution', see Pearse, and on **Africa**, key sources are Ford, Paul Harrison, Richards, also Stern.

1 J. Boyd-Orr, *As I Recall*, London: MacGibbon and Kee, 1966.
2 M. G. McGarry and J. Stainforth (eds), *Compost, Fertilizer, and Biogas Production from Human Wastes in the People's Republic of China*, Ottawa: International Development Research Centre, 1978.
3 M. Wilson and L. Thompson, *The Oxford History of South Africa*, volume I, London: Oxford University Press, 1969, p. 132.
4 J. Ford, *The Role of Trypanosomiasis in African Ecology*, Oxford: Clarendon Press, 1971, p. 8, and on Moçambique, pp. 332–9.
5 P. Richards, *Indigenous Agricultural Revolution*, 1985, p. 10.
6 G. Benneh, 'Technology should seek tradition', *Ceres (FAO Review)*, volume III, no. 5 (September/October), 1970.
7 D. M. Maurya, A. Bottrell, and J. Farrington, 'Improved livelihoods, genetic diversity, and farmer participation', *Experimental Agriculture*, 24 (1985), pp. 311–20.
8 Liu Chenlie, 'Now they're cooking with gas', in *Energy Resources in the Countryside*, Beijing: China Reconstructs reprints, 1984; see also McGarry and Stainforth, *Compost, Fertilizer, and Biogas Production*, 1978.
9 Richards, *Indigenous Agricultural Revolution*, 1985, p. 145.
10 D. Rocheleau and L. Malaret, 'Ethnoecological methods and farmer innovation in agroforestry', paper presented at a workshop on Farmers and Agricultural Research, University of Sussex, Brighton, July 1987. A revised version appears in R. Chambers, A. Pacey, and L.-A. Thrupp (eds), *Farmer First: Farmer Innovation and Agricultural Research*, 1989.
11 P. Stern, *Small-Scale Irrigation*, 1979, p. 13. Other material on African water conservation techniques is given by N. W. Hudson, *Soil and Water Conservation in Semi-arid Areas*, Rome: Food and Agriculture Organization, 1987; also W. T. W. Morgan, 'Sorghum gardens in South Turkana', *Geographical Journal*, 140 (1970), pp. 80–93. For examples discussed by archaeologists, see D. Randall-MacIver, *Medieval Rhodesia*, London 1906, reprinted London: Cass, 1971.
12 C. Juma, 'Ecological complexity and agricultural innovation', paper delivered at a workshop on Farmers and Agricultural Research, University of Sussex, Brighton, July 1987, drawing on Juma's *Genetic Resources and Biotechnology in Kenya*, Nairobi: Public Law Institute, 1987.
13 A. Pacey and A. Cullis, *Rainwater Harvesting*, 1986, p. 62.
14 M. Kinyanjui, 'The jiko industry in Kenya', in Robin Clarke (ed.), *Wood-*

stove Dissemination, London: Intermediate Technology Publications, 1985; also P. Harrison, *The Greening of Africa*, London: Paladin, 1987, pp. 211–16.

15 *A Guide to the Safe Storage of Cereals, Oilseeds and Pulses*, Lilongwe (Malawi): Extension Aids Branch, Ministry of Agriculture, 1973.

16 S. B. Watt, *Ferrocement Water Tanks and their Construction*, London: Intermediate Technology Publications, 1978, pp. 84–91.

17 Harrison, *The Greening of Africa*, 1987, pp. 165–70.

18 Richards, *Indigenous Agricultural Revolution*, 1985, pp. 146–9.

19 Ibid., pp. 70–1, 84.

20 Ibid., p. 58.

21 On alley cropping, see Harrison, *The Greening of Africa*, 1987, pp. 192–9; also J. Sumberg and C. Okali, 'Farmers, on-farm research, and the development of new technology', *Experimental Agriculture*, 24 (1988), pp. 333–42 (and note that the journal *Plant and Soil* published B. T. Kang's work in 1981; see volume 63, pp. 165–79).

22 Qaisar, *The Indian Response to European Technology and Culture, 1498–1707*, 1982, pp. 23–6.

23 Jeon, *Science and Technology in Korea*, 1974.

24 T. G. Rawski, *American Economic Review*, 65 (1975), no. 2 (Papers and Proceedings), pp. 383–96.

25 R. McCarrison, 'Faulty food in relation to gastro-intestinal disorder', *Journal of the American Medical Association*, 78 (1922), pp. 1–8; also R. McCarrison and B. Viswanath, 'The effect of manurial conditions on vitamin values of millet and wheat', *Indian Journal of Medical Research*, 14 (1926–7), pp. 351–78.

26 Needham, *Science and Civilization in China*, volume I, 1954, p. 149.

27 Alvares, *Homo Faber: technology and culture in India, China and the West*, 1980, foreword by Rajni Kothari, p. (xi).

28 Mikhail Gorbachev, address to the United Nations General Assembly, as reported in the *Guardian*, 8 December 1988.

Bibliography

This bibliography gives details of three categories of source material: books referred to by author's name only in captions or notes on illustrations and tables; those referred to by author's name only in the comments on background material; and those of general importance for the subject.

Ajayi, J. F. A. and Crowder, Michael (eds), *History of West Africa*, volume I. London: Longman, 1971.

al-Hassan, *see* Hassan.

Alvares, Claude A., *Homo Faber: technology and culture in India, China and the West from 1500 to the present day*. The Hague: Martinus Nijhoff, 1980.

Atil, Esin, *The Age of Sultan Suleyman the Magnificent*. Washington (DC): National Gallery of Art, and New York: Abrams, 1987.

Bates, L. F., *Sir Alfred Ewing: a pioneer in physics and engineering*. London: Longman (for the British Council), 1946.

Bhattacharya, S., and Chaudhuri, B., 'The regional economy . . . Eastern India', in D. Kuma and M. Desai (eds), *Cambridge Economic History of India*, volume II. Cambridge: Cambridge University Press, 1983.

Born, W., 'The Indian hand-spinning wheel and its migration to east and west', *CIBA Review* (Basle), no. 28, December 1939.

Boserup, Ester, *Population and Technology*. Oxford: Blackwell, 1981 (also published as *Population and Technological Change*. Chicago: Chicago University Press, 1981).

Bovill, E. W., *The Gold Trade of the Moors*. London: Oxford University Press, 1968.

Bradfield, M., *The Changing Pattern of Hopei Agriculture*. London: Royal Anthropological Institute, 1971.

Braudel, F., *Civilization and Capitalism: Volume 3, The Perspective of the World*. London: Collins, 1984, trans. Sian Reynolds.

Bray, Francesca, *Agriculture*, in J. Needham (ed.), *Science and Civilization in China*, volume VI, part 2. Cambridge: Cambridge University Press, 1984.

Bray, Francesca, *The Rice Economies: technology and development in Asian societies*. Oxford: Blackwell, 1986.

Bronson, B., 'Terrestrial and meteoritic nickel in Indonesian kris', *Historical Metallurgy*, 21 (1), 1987, pp. 8–15.

Bühler, Alfred, 'The geographical extent of the use of bark fabrics', *CIBA Review* (Basle), no. 33, 1940, pp. 1,170–9.

Cardwell, D. S. L., *The Organization of Science in England*. London: Heinemann, 1957.
Carman, W. Y., *A History of Firearms from Earliest Times to 1914*. London: Routledge, 1955.
Chaloner, W. H., 'Sir Thomas Lombe (1685–1739) and the British silk industry', *History Today*, 3 (1953), pp. 778–85.
Chamberlain, W. H., *Japan over Asia*. London: Duckworth, 1938.
Chambers, R., Pacey, A., and Thrupp, L.-A. (eds), *Farmer First: Farmer Innovation and Agricultural Research*. London: Intermediate Technology Publications, 1989.
Chaudhuri, K. N., *The Trading World of Asia and the English East India Company*. Cambridge: Cambridge University Press, 1975.
Chibbett, David, *The History of Japanese Printing and Book Illustration*. Tokyo: Kodansha International, 1977.
Cipolla, Carlo M., *Guns and Sails*. London: Collins, 1965.
Cipolla, Carlo M., *European Culture and Overseas Expansion* (incorporating *Guns and Sails* [1965] and *Clocks and Culture* [1967] without their illustrations). Harmondsworth: Penguin, 1970.

Daumas, Maurice (ed.), *Histoire générale des techniques*. 5 volumes, Paris: Presses Universitaires de France, 1962–79.
Dharampal, *Indian Science and Technology in the Eighteenth Century*. Delhi: Impex India, 1971.
Dunsheath, P., *A History of Electrical Engineering*. London: Faber, 1962.
Dutt, Romesh, *The Economic History of India under early British rule*. London, 1901; 7th edn, London, Routledge and Kegan Paul, 1950.
Dutt, Romesh, *The Economic History of India in the Victorian Age*. London, 1903; 7th edn, London, Routledge and Kegan Paul, 1950.

Egerton, Lord, *Indian and Oriental Armour*. New edn., 1896, reprinted London: Arms and Armour Press, 1968.
Elvin, Mark, *The Pattern of the Chinese Past*. London: Eyre Methuen, 1973.
Esper, Thomas, 'Industrial serfdom and metallurgical technology in nineteenth century Russia', *Technology and Culture*, 23 (1982), pp. 583–608.
Ewell, P. T., and Merrill-Sands, D., 'Milpa in Yucatan', in B. L. Turner and S. B. Brush (eds), *Comparative Farming Systems*. New York: Guilford Publications, 1987.

Farrar, D. M., and Pacey, A. J., 'Aspects of the German tradition in technical education', in D. S. L. Cardwell (ed.), *Artisan to Graduate*. Manchester: Manchester University Press, 1974.
Fei, J. C. H., and Ranis, G., 'Innovation, capital accumulation and economic development (in Japan)', *American Economic Review*, 53, (1963), pp. 283–313, and 54 (1964), pp. 1,063–8.

Feuerwerker, Albert, *China's Early Industrialization*. Cambridge (Mass.), Harvard University Press, 1958, reprinted New York: Atheneum, 1970.

Ford, John, *The Role of Trypanosomiasis in African Ecology*. Oxford: Clarendon Press, 1971.

Gibbs-Smith, C. H., *The Aeroplane: an historical survey*. London: Her Majesty's Stationery Office, 1960.

Gille, Bertrand, *The History of Techniques: Volume I, Techniques and Civilizations*. Montreux (Switzerland): Gordon and Breach, 1986.

Gimpel, Jean, *The Medieval Machine: the industrial revolution of the Middle Ages*. London: Gollancz, 1977.

Gittinger, Mattiebelle, *Master Dyers to the World*. Washington (DC): The Textile Museum, 1982.

Goonatilake, S., *Aborted Discovery: science and creativity in the Third World*. London: Zed Books, 1984.

Habakkuk, H. J., *American and British Technology in the Nineteenth Century*. Cambridge: Cambridge University Press, 1962.

Habib, Irfan, 'Technological changes and society', in D. Chattopadhyaya (ed.), *Studies in the History of Science in India*, volume II. Delhi: Editorial Enterprises, 1982.

Harrison, Gordon, *Mosquitoes, Malaria and Man*. London: John Murray, 1978.

Harrison, Paul, *The Greening of Africa*. London: Paladin, 1987.

Harrison, P. D., and Turner, B. L. (eds), *Pre-historic Maya Agriculture*. Albuquerque: University of New Mexico Press, 1978.

Harriss, Joseph, *The Eiffel Tower: symbol of an age*. London: Paul Elek, 1976.

Hartwell, Robert, 'Markets, technology and the structure of enterprise in the eleventh century Chinese iron and steel industry', *Journal of Economic History*, 26 (1966), pp. 29–58.

al-Hassan, A. Y., and Hill, D. R., *Islamic Technology: an illustrated history*. Paris: UNESCO, and Cambridge: Cambridge University Press, 1986.

Headrick, D. R., *The Tools of Empire: technology and European imperialism in the nineteenth century*. New York: Oxford University Press, 1981.

Hill, Donald, *A History of Engineering in Classical and Medieval Times*. Beckenham (England): Croom Helm, 1984.

Hills, R. L., *Power in the Industrial Revolution*. Manchester: Manchester University Press, 1970.

Hills, R. L., *Richard Arkwright and Cotton Spinning*. London: Priory Press, 1973.

Ho Ping-ti, *Studies on the Population of China, 1368–1953*. Cambridge (Mass.): Harvard University Press, 1959.

Hobsbawm, E. J., *Industry and Empire: an economic history of Britain since 1750*. London: Weidenfeld and Nicolson, 1968.

Hopkins, H. J., *A Span of Bridges*. Newton Abbot (England): David and Charles, 1970.

Horridge, G. A., *The Design of Planked Boats of the Moluccas*. London: National Maritime Museum, 1978.

Horridge, G. A., *The Lashed-lug Boat of the Eastern Archipelagoes*. London: National Maritime Museum, 1982.

Hua Jueming, 'Mass production of iron castings in ancient China', *Scientific American*, 248 (1), 1983, pp. 120–8.

Hunter, Louis, *Steamboats on the Western Rivers*. Cambridge (Mass.): Harvard University Press, 1949.

Hyde, Charles K., *Technological Change and the British Iron Industry*. Princeton (NJ): Princeton University Press, 1977.

Ikram, S. M., *Muslim Civilization in India*. New York: Columbia University Press, 1964.

Irwin, John, and Brett, Katharine, *Origins of Chintz*. London: Her Majesty's Stationery Office, 1970.

Issawi, Charles, *The Economic History of Turkey, 1800–1914*. Chicago: Chicago University Press, 1980.

Jeon, Sang-woon, *Science and Technology in Korea*. Cambridge (Mass.): MIT Press, 1974.

Jones, H. J., *Live Machines: hired foreigners and Meiji Japan*. Vancouver: University of British Columbia, 1980.

Juvet-Michel, A., 'Textile printing in eighteenth century France', *CIBA Review* (Basle), no. 31, March 1940, pp. 1,091–9.

Kleinman, A., Kunstadter, P., Alexander, E. R., and Gale, J. L. (eds), *Medicine in Chinese Cultures: comparative studies of health care in Chinese and other societies*. Washington (DC): US Department of Health, Education and Welfare, 1975.

Klemm, Friedrich, *A History of Western Technology*. London: Allen & Unwin, 1959.

Konczacki, Z. A., and Konczacki, J. M. (eds), *An Economic History of Tropical Africa*. London: Cass, 1977.

Kubata, Kurao, *Tatara Process Iron-making in Japan* (in Japanese). Tokyo: Komine Shoten Publishing Co., 1982.

Kumar, Dhama, and Desai, Meghnad (eds), *Cambridge Economic History of India, Volume II, 1757–1970*. Cambridge: Cambridge University Press, 1983.

Lach, Donald F., *Asia in the Making of Europe*, volumes I and II. Chicago: Chicago University Press, 1977.

Landes, David S., *The Unbound Prometheus: technological change and industrial development in western Europe*. Cambridge: Cambridge University Press, 1969.

Langrish, J., 'The changing relationship between science and technology', *Nature* (London), 250 (1974), pp. 64–5.

Liang, Ernest P., *China: railways and agricultural development*. Chicago: University of Chicago Department of Geography, 1982.

Littlefield, D. C., *Rice and Slaves*. Baton Rouge and London: Louisiana State University Press, 1981.

Liu Guojun and Zheng Rusi, *The Story of Chinese Books*. Beijing: Foreign Language Press, 1985.

Lockwood, W. W., *The Economic Development of Japan: growth and structural change, 1868–1938*. Princeton (NJ): Princeton University Press, 1954, and London: Oxford University Press, 1955; expanded edn. 1968.

McGarry, M. G., and Stainforth, J. (eds), *Compost, Fertilizer, and Biogas Production from Human Wastes in the People's Republic of China*. Ottawa: International Development Research Centre, 1978.

McKeown, Thomas, *The Role of Medicine: dream, mirage or nemesis?*. Oxford: Blackwell, 1966.

McNeill, W. H., *Plagues and People*. Oxford: Blackwell, 1977.

McNeill, W. H., *The Pursuit of Power: technology, armed force and society*. Chicago: Chicago University Press, 1982, and Oxford: Blackwell, 1983.

Macpherson, W. J., *The Economic Development of Japan, 1868–1941*. Basingstoke (England): Macmillan, 1987.

Mao Yi-sheng, *Bridges in China, Old and New*. Peking (Beijing): Foreign Language Press, 1978.

Marcus, Joyce, 'The plant world of the sixteenth and seventeenth century lowland Maya', in K. V. Flannery (ed.), *Maya Subsistence*. New York: Academic Press, 1982.

Marx, Leo, *The Machine in the Garden: technology and the pastoral ideal in America*. New York: Oxford University Press, 1964.

Matheson, Ewing, *Works in Iron*. London, 1877.

Mathur, Krishnan D., 'Indian astronomy in the era of Copernicus', *Nature*, 251 (1974), pp. 283–5.

Mitchell, B. R., 'Statistical appendix, 1700–1914', in Carlo M. Cipolla (ed.), *The Fontana Economic History of Europe*, volume IV, part 2. London: Fontana/Collins, 1973.

Mookerji, Radhakumud, *Indian Shipping: a history of sea-borne trade and maritime activity*. Bombay: Longman Green, 1912.

Morgan, David, *The Mongols*. Oxford: Blackwell, 1986.

Morison, Elting E., *From Know-How to Nowhere: the development of American technology*. New York: Basic Books, and Oxford: Blackwell, 1974.

Needham, Joseph, *Science and Civilization in China*, 6 volumes. Cambridge: Cambridge University Press, 1954 to date. See especially: volume IV, part 2 (1965, co-author Wang Ling), *Mechanical Engineering*; volume IV, part 3 (1971, co-authors Wang Ling and Lu Gwei-Djen), *Civil Engineering and Nautics*; volume V, part 1 (1985, sole author Tsien Tsuen-Hsuin), *Paper and Printing*; volume V, part 7 (1986, co-authors Ho Ping-Yu, Lu Gwei-Djen and Wang Ling), *Military Technology: the gunpowder epic*; volume VI, part 2 (1984, sole author Francesca Bray), *Agriculture*.

Needham, Joseph, *The Development of Iron and Steel Technology in China*. London: Newcomen Society (Dickinson Memorial Lecture), 1958.

Needham, Joseph, Wang Ling, and Price, Derek J. de Solla, *Heavenly Clockwork: the great astronomical clocks of medieval China*. Cambridge: Cambridge University Press, 1960.

Niñez, Vera K., *Household Gardens: theoretical considerations on an old survival strategy*. Lima (Peru): International Potato Centre, 1984.

Niñez, Vera K. (ed.), *Household Food Production: comparative perspectives*. Lima (Peru): International Potato Centre, 1985.

North, Anthony, *An Introduction to Islamic Arms*. London: Her Majesty's Stationery Office, 1985.

Okita, S., 'Choice of technique – Japan's experience', in Kenneth Berrill (ed.), *Economic Development with special reference to East Asia*. London: Macmillan, 1966.

Oliver, Roland, and Fage, J. D., *A Short History of Africa*. Harmondsworth: Penguin (1962), 3rd edn 1970.

Pacey, A., *The Maze of Ingenuity: ideas and idealism in the development of technology*. Cambridge (Mass.): MIT Press, 1976.

Pacey, A., *The Culture of Technology*. Cambridge (Mass.): MIT Press, and Oxford: Blackwell, 1983.

Pacey, A., and Cullis, A., *Rainwater Harvesting*. London: Intermediate Technology Publications, 1986.

Pant, G. N., *Studies in Indian Weapons and Warfare*. Delhi: S. J. Singh, 1970.

Pearse, Andrew, *Seeds of Plenty, Seeds of Want: social and economic implications of the green revolution*. Oxford: Clarendon Press, 1980.

Pringle, Peter, and Spigelman, James, *The Nuclear Barons*. New York: Holt, Rinehart and Winston, 1981.

Qaisar, A. J., *The Indian Response to European Technology and Culture, 1498–1707*. Delhi: Oxford University Press, 1982.

Ranis, Gustav, 'Factor proportion in Japanese economic development', *American Economic Review*, 47 (1957), pp. 594–607.

Rawski, Thomas G., 'Problems of technology absorption in Chinese industry', *American Economic Review*, 65 (1975), no. 2 (Papers and Proceedings), pp. 383–96.

Ray, P. (ed.), *History of Chemistry in Ancient and Medieval India*, edited from the work of P. C. Ray. Calcutta: Indian Chemical Society, 1956.

Raychaudhuri, T., 'The mid-eighteenth century background', in D. Kumar and M. Desai (eds), *Cambridge Economic History of India, Volume II, 1757–1970*. Cambridge: Cambridge University Press, 1983.

Raychaudhuri, Tapan, and Habib, Irfan (eds), *Cambridge Economic History of India, Volume I, 1200–1750*. Cambridge: Cambridge University Press, 1982.

Reij, C., Turner, S., and Kuhlman, T., *Soil and Water Conservation in Sub-Saharan Africa*. Amsterdam: Centre for Development Co-operation, Free University, 1986.

Richards, Paul, *Indigenous Agricultural Revolution: ecology and food production in West Africa*. London: Hutchinson, 1985.

Rosenberg, Nathan, *Technology and American Economic Growth*. New York: Harper and Row, 1972.

Rosenberg, Nathan, *Perspectives on Technology*. Cambridge: Cambridge University Press, 1976.

Rostoker, W., Bronson, B., and Dvorak, J., 'The cast iron bells of China', *Technology and Culture*, 25 (1984), pp. 750–67.

Roth, H. Ling, *Studies in Primitive Looms*. Halifax (England), 1918; reprinted Bedford (England): Ruth Bean, 1977.

Sansom, George, *A History of Japan, 1334–1615*. London: Cresset Press, 1961.

Sarkar, J. N., *The Art of War in Medieval India*. Delhi: Munshiram Manoharlal Publishers, 1984.

Schaefer, Gustav, 'Madder and Turkey Red', *CIBA Review* (Basle), no. 39, May 1941.

Schaefer, Gustav, and Haller, R., 'Medieval cloth printing in Europe', *CIBA Review* (Basle), no. 26, October 1939.

Scott, J. D., *Siemens Brothers, 1858–1958: an essay in the history of industry*. London: Weidenfeld and Nicolson, 1958.

Simkin, C. G. F., *The Traditional Trade of Asia*. London: Oxford University Press, 1968.

Singer, C., Holmyard, E. J., Hall, A. R., and Williams, T. I. (eds), *A History of Technology*, 7 volumes. London: Oxford University Press, 1954–9 and 1978. (Volumes 6 and 7 ed. T. I. Williams only.)

Smith, Cyril Stanley, *A History of Metallography*. Chicago: Chicago University Press, 1960.

Smith, Norman, *A History of Dams*. London: Peter Davies, 1971.

Stern, Peter, *Small-Scale Irrigation*. London: Intermediate Technology Publications, 1979.

Summers, Roger, *Ancient Mining in Rhodesia and Adjacent Areas*. Salisbury (now Harare): National Museums of Rhodesia, Memoir no. 3, 1969.

Sutton, J. E. G., *Early Trade in Eastern Africa*. Historical Association of Tanzania, Paper no. 11. Nairobi: East African Publishing House, 1973.

Sykes, Peter, 'From mauve to macromolecules', in A. and N. Clow (eds), *Science News 41*. Harmondsworth: Penguin, 1956.

Thomson, John, *China, the Land and its People*. London, 1873; reprinted, Hong Kong: John Warner Publications, 1977.

Thorburn, Craig, *Teknologi Kampungan: indigenous Indonesian technologies*. Stanford (Calif.): Volunteers in Asia, 1982.

Trinder, Barrie, *The Industrial Revolution in Shropshire*. Chichester (England): Phillimore, 1973.

Tsien Tsuen-Hsuin, *Paper and Printing*, in J. Needham (ed.), *Science and Civilization in China*, volume V, part 1, Cambridge: Cambridge University Press, 1985.

Wadia, Ruttonjee Ardeshir, *The Bombay Dockyard and the Wadia Master Builders*. Bombay: R. A. Wadia, 1957.

Wadsworth, A. P., and Mann, J. de L., *The Cotton Trade and Industrial Lancashire*. Manchester: Manchester University Press, 1931.

Wagner, Donald B., 'Some traditional Chinese iron production techniques practised in the twentieth century', *Historical Metallurgy*, 18 (1984), pp. 95–104.

Watson, Andrew M., *Agricultural Innovation in the Early Islamic World.* Cambridge: Cambridge University Press, 1983.

Weller, Jac, *Wellington in India.* London: Longman, 1972.

Westwood, J. N., *A History of Russian Railways.* London: Allen and Unwin, 1964.

White, Lynn, *Medieval Technology and Social Change.* London: Oxford University Press, 1962.

Wilkinson, Richard, *Poverty and Progress.* London: Methuen, 1973.

Willetts, William, *Chinese Art*, 2 volumes. Harmondsworth: Penguin, 1958.

Williams, T. I., *The Chemical Industry Past and Present.* Harmondsworth: Penguin, 1953.

Wittfogel, Karl A., *Oriental Despotism: a comparative study of total power.* New Haven: Yale University Press, 1957.

Worsley, Peter, *The Three Worlds: culture and world development.* London: Weidenfeld and Nicolson, 1984.

Wulff, Hans E., *The Traditional Crafts of Persia.* Cambridge (Mass.): MIT Press, 1966.

Index